SpringerBriefs in Applied Sciences and Technology

PoliMI SpringerBriefs

Editorial Board

Barbara Pernici, Politecnico di Milano, Milano, Italy

Stefano Della Torre, Politecnico di Milano, Milano, Italy

Bianca M. Colosimo, Politecnico di Milano, Milano, Italy

Tiziano Faravelli, Politecnico di Milano, Milano, Italy

Roberto Paolucci, Politecnico di Milano, Milano, Italy

Silvia Piardi, Politecnico di Milano, Milano, Italy

More information about this subseries at http://www.springer.com/series/11159
http://www.polimi.it

Angelo Geraci

Editor

Special Topics in Information Technology

POLITECNICO
MILANO 1863

Editor
Angelo Geraci
Dipartimento di Elettronica
Informazione e Bioingegneria (DEIB)
Politecnico di Milano
Milan, Italy

ISSN 2191-530X   ISSN 2191-5318   (electronic)
SpringerBriefs in Applied Sciences and Technology
ISSN 2282-2577   ISSN 2282-2585   (electronic)
PoliMI SpringerBriefs
ISBN 978-3-030-62475-0   ISBN 978-3-030-62476-7   (eBook)
https://doi.org/10.1007/978-3-030-62476-7

© The Editor(s) (if applicable) and The Author(s) 2021. This book is an open access publication.

Open Access This book is licensed under the terms of the Creative Commons Attribution 4.0 International License (http://creativecommons.org/licenses/by/4.0/), which permits use, sharing, adaptation, distribution and reproduction in any medium or format, as long as you give appropriate credit to the original author(s) and the source, provide a link to the Creative Commons license and indicate if changes were made.

The images or other third party material in this book are included in the book's Creative Commons license, unless indicated otherwise in a credit line to the material. If material is not included in the book's Creative Commons license and your intended use is not permitted by statutory regulation or exceeds the permitted use, you will need to obtain permission directly from the copyright holder.

The use of general descriptive names, registered names, trademarks, service marks, etc. in this publication does not imply, even in the absence of a specific statement, that such names are exempt from the relevant protective laws and regulations and therefore free for general use.

The publisher, the authors and the editors are safe to assume that the advice and information in this book are believed to be true and accurate at the date of publication. Neither the publisher nor the authors or the editors give a warranty, expressed or implied, with respect to the material contained herein or for any errors or omissions that may have been made. The publisher remains neutral with regard to jurisdictional claims in published maps and institutional affiliations.

This Springer imprint is published by the registered company Springer Nature Switzerland AG
The registered company address is: Gewerbestrasse 11, 6330 Cham, Switzerland

Preface

In the frame of doctoral studies in Information Technology at the Department of Electronics, Information and Bioengineering at the Politecnico di Milano, this is a compilation of the most significant research projects whose theses were defended in 2020 and selected for the IT Ph.D. Award. The Ph.D. in Information Technology is characterized by a strong interdisciplinary value in the IT sector which is divided into four main research lines, Telecommunications, Electronics, Computer Science and Engineering, and Systems and Control. Each of these areas develops state-of-the-art research that often proves to be enabling the evolution of innovative systems in the IT field. The coexistence of such orthogonal skills proves to be an excellent catalyst for collaboration between young researchers who are able to generate significant synergies between researches, also developing strong integration skills in open and dynamic research groups. The aim of this collection is to provide a cutting-edge overview of the most recent research trends in Information Technology developed at the Politecnico di Milano, in an easy-to-read format to present the main results even to non-specialists in the specific field.

Milan, Italy   Angelo Geraci
July 2020

Contents

Part I
Telecommunications

Chapter 1
Machine-Learning Defined Networking: Towards Intelligent Networks

Sebastian Troia

1.1 Introduction

On October 29, 1969, the world was changed forever. At 10 : 30 p.m., a student programmer at the University of California at Los Angeles (UCLA) named Charley Kline sent the letter "l" and the letter "o" electronically more than 350 miles to a Stanford Research Institute computer in Menlo Park, California. In that specific moment the Internet history was made and a technological revolution had begun. 26 years after Charley Kline's first message, the number of people interconnected by Internet was 16 million and in just 50 years it reached 4 billion (58.8% of world population).[1] Bit streams of information over the Internet Protocol (IP), known as IP traffic, continuously crosses the world to reach: end users through High Definition (HD) screens (either in mobile and fixed devices), connected machines (Internet of Things), and IT infrastructures, i.e., public or private datacenters and the Cloud infrastructures. IP traffic is powered by ubiquitous Internet connectivity provided by CSPs. Every day, around 18 billion Internet-connected devices and 4 billion global Internet users transfer 900 Petabytes (10^{15} bytes) of IP traffic across the globe [1]; by the end of 2021, the global IP traffic will reach 3.3 zettabytes (10^{21} bytes) [2]. Moreover, with the advent of 5G technology, we are witnessing the development of increasingly bandwidth-hungry network applications, such as enhanced mobile broadband, massive machine-type communications and ultra-reliable low-latency communications. Contrary to the evolution of previous generations, 5G will require not only improved networking solutions but also more sophisticated mechanisms for traffic management to fulfill stronger end-to-end Quality of Service (QoS) require-

---

[1]Online: https://www.internetworldstats.com/stats.htm.

S. Troia (✉)
Dipartimento di Elettronica, Informazione e Bioingegneria, Politecnico di Milano,
Via Giuseppe Ponzio, 34, 20133 Milano, MI, Italy
e-mail: sebastian.troia@polimi.it

© The Author(s) 2021
A. Geraci (ed.), *Special Topics in Information Technology*,
PoliMI SpringerBriefs, https://doi.org/10.1007/978-3-030-62476-7_1

3

ments of numerous different "verticals", such as automotive, manufacturing, energy, eHealth, agriculture, etc.

Traditional network architectures are ill-suited to meet the requirements of the aforementioned emerging Internet services. The common practice in CSP networks is to perform a static resource allocation which meets the peak-hour traffic demand. This method leads to poor network and energy efficiency, as outside the peak hour resources will be over-provisioned. As a consequence, the Internet has become one of the major energy consumer in the world. This situation is forcing the CSPs to change the underlying network technologies, and have started to look at new technological solutions that increase the level of programmability, control, and flexibility of configuration, while reducing the overall costs related to network operations.

This Ph.D. thesis [3] focuses on three of the major disruptive technologies in networking in the last two decades: Software Defined Networking (SDN), Network Function Virtualization (NFV) and Machine Learning (ML).

SDN is one of the fastest growing markets in Information and Communication Technologies (ICT), from 289 million in 2015 to 8:7 billion in 2020, with a Compound Annual Growth Rate (CAGR) of 98% [4]. It combines automation, agility and development of applications to deliver network services in a deterministic, dynamic, and scalable manner. By separation of control and data plane, SDN brings networking back to centralized control, allowing applications to program the network forwarding rules through logically centralized controllers. SDN brings a promising solution to introduce network programmability and accelerate networking innovation. For instance, network programmability allows to implement intelligent dynamic optimization methods to improve performance and energy efficiency. The application of SDN in Enterprise Networking (EN) is called Software Defined Wide Area Network (SD-WAN) [5] and represents an emerging paradigm that introduces the advantages of SDN into EN. SD-WAN is a revolutionary way to simplify enterprise networks and achieve optimal application performance using centrally managed WAN virtualization. Unlike traditional WAN connections, SD-WAN provides high network agility and cost savings.

By leveraging virtualization technologies, the European Telecommunications Standards Institute (ETSI) proposed NFV to virtualize network services previously performed by dedicated proprietary hardware. NFV is the paradigm of transferring network functions from dedicated hardware appliances to software-based applications running on commercial off-the-shelf equipment [6]. Network devices are no longer proprietary but are open to host software from different vendors, drastically reducing CapEx and OpEx. Although the adoption of SDN and NFV in networking brings revolutionary benefits in scale and agility, it also brings a whole new level of complexity. Virtualization breaks traditional networking into dynamic components and layers that have to work in unison and that can change at any given time [7], for instance, a virtualized firewall can be subject to continuous updates by the CSP. The high complexity introduced by these new technologies has led research to focus on increasingly intelligent methods and algorithms to optimize the allocation of resources in all network segments: access, metro and core. In recent years, Machine Learning has become the main mathematical tool used by researchers in this field. ML is a sub-field of computer science that originated from the study of pattern recognition and computational learning theory in artificial intelligence. By leveraging complex mathematical and statistical tools, computers can learn by input data how to solve specific problems that have been traditionally solved by human beings [8]. ML explores algorithms that can "learn" from and make predictions out of input data. The learning phase is also referred to as "training" an algorithm, by iterative and feedback-based mechanisms, to solve specific problems. Such algorithms operate by building a model in order to make data-driven predictions or decisions, rather than following static program instructions. ML algorithms are typically classified into three different categories:

- **Supervised Learning (SL)** is used in a variety of applications, such as speech recognition, spam detection and object recognition. The goal is to predict the value of one or more output variables given the value of a vector of input variables. The output variable can be a continuous variable (regression problem) or a discrete variable (classification problem). A training data set comprises a set of samples of the input variables and the corresponding output values.
- **Unsupervised Learning (UL)** consists of training an algorithm using only input vector data with the aim to find hidden patterns and similarities. Social network analysis, genes clustering and market research are among the most successful applications of unsupervised learning methods. It can address different tasks, such as clustering or cluster analysis. Clustering is the process of grouping data so that the intra-cluster similarity among the input data samples is high, while the inter-cluster similarity is low.
- **Reinforcement Learning (RL)** is an approach to ML that trains algorithms using a system of positive and negative rewards. Contrary to what happens with the supervised and unsupervised learning, in RL we do not have training, validation and test set of data defined "a priori". An RL algorithm, or *agent*, learns how to perform a task by interacting with its *environment*. The interaction is defined in terms of specific *actions*, *observations* and *rewards*.

This chapter presents the main results of my PhD thesis [3] by introducing novel ML-based algorithms to solve various network optimization problems. The thesis provides contributions to all the ML branches by implementing supervised, unsupervised and reinforcement learning algorithms together with their implementations in real and emulated SDN/NFV testbed scenarios.

1.2 Network Traffic Prediction

Mobile, metro and core networks often suffer from resource inefficiency due to over-provisioning. It's a common practice to perform static resource allocation based on the peak-hour demand, because the current operational processes used by the network operators are too slow to dynamically allocate the resources following the daily demand variations. Over-provisioning leads to poor energy efficiency and high OpEx, as the resources are sub-utilized outside of the peak hour. Moreover, as the peak-hour to average demand ratio continue to increase [2], the static resource allocation lead to higher and unnecessary OpEx and CapEx. As such, understanding how to correctly predict network behaviour plays a vital role in the management and provisioning of mobile and fixed network services. Network traffic prediction has become very important for network providers. Having an accurate traffic prediction tool is essential for most network management tasks, such as resource allocation, short-time traffic scheduling or re-routing, long-term capacity planning, network design and network anomaly detection. A proactive prediction-based approach allows network providers to optimize network resource allocation, potentially improving the QoS. With this aim, we introduce a matheuristic (i.e. interoperation of heuristic and mathematical programming) for dynamic network optimization. In Fig. 1.1 we show the proposed model. Traffic is predicted on different spatial locations developing different machine-learning algorithms, such as Artificial Neural Network (ANN) [9], Recurrent Neural Network (RNN) [10] and Diffusion Convolutional RNN (DCRNN) [11]. The predicted traffic-demand is then used to optimize the network at various hours during the day, so to adapt resource occupation to the actual traffic volume. The optimization problem solvers are based on Integer Linear Programming (ILP) and heuristic algorithms; their goal is to minimize the power and energy consumption of the network elements, providing weights to assign to network links every hour. After that, a weighted graph is generated to route the actual traffic demands by a minimum-cost path algorithm. The time required by optimization is not an issue, since prediction allows to start performing optimization computation in advance (Offline). Such feature makes the solution reported suitable to be implemented as an SDN application for 5G scenarios. We have chosen the metro infrastructure of a mobile operator as use case; in particular, the objective of the resource optimization is the optical metro network used as the backbone for the mobile service. The tests

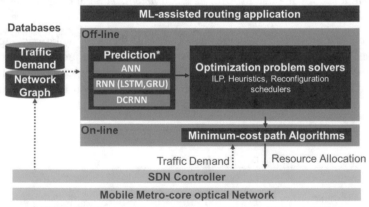

* **ANN**: Artificial Neural Network; **RNN**: Recurrent Neural Network; **LSTM**: Long Short Term Memory Network; **GRU**: Gated Recurrent Unit; **DCRNN**: Diffusion Convolutional RNN;

Fig. 1.1 Dynamic optical routing matheuristic for software-defined mobile metro network [9]

demonstrate the effectiveness of our methodology, with results that match almost per-
fectly th!htbe behaviour of a network that performs optical routing reconfiguration
with a perfect, oracle-like traffic prediction. The methodology and the detailed results
can be found in Chaps. 2 and 3 of my Ph.D. thesis and in the following papers: [9–12].

1.3 Network Traffic Pattern Identification

Due to the highly predictable daily movements of citizens in urban areas, mobile
traffic shows repetitive patterns with spatio-temporal variations. This phenomenon
is known as tidal effect analogy to the rise and fall of the sea levels. Recognizing
and defining traffic load patterns at the base station thus plays a vital role in traffic engineering, network design and load balancing since it represents an important
solution for the Internet Service Providers (ISPs) that face network congestion problems or over-provisioning of the link capacity. State of the art works have dealt with
the classification and identification of patterns through the use of techniques, which
inspect the flow of data of a particular application. In the previous section we have
shown dynamic network optimization techniques based on supervised learning algorithms, or prediction algorithms, while in this section, we use unsupervised learning to
"learn" periodic trends (or patterns) from traffic data over time. We wanted to answer
to the following questions: Do different areas of the city with the same social function
display the same typical traffic pattern? Can they be used to optimize the metro-core
optical network? In order to answer to these questions we analysed datasets concerning the city of Milan [13, 14]. They show information regarding the internet traffic of
Telecom Italia (TIM) service provider and the point of interests located throughout
the city, such as schools, restaurants, social services, etc., that reveal the social function of a particular area. We developed a novel model for basic pattern identification
based on matrix factorization methods by integrating the aforementioned datasets in
order to assess the similarity of traffic patterns related to areas with the same type of
point of interests. In the field of pattern recognition, the Non-negative Matrix Factorization (NMF) is one of the most used method thanks to the ability to detect basic
flows inside large matrices. First, we apply the classical NMF method to a real-world
dataset that collects the data traffic of mobile users at the base station level in the
city of Milan. Afterwords, we propose an integration of multi-domain data on the
basis of NMF and denote it as Collective-NMF based model. The final results show
the presence of very similar patterns located in distant areas of the city that share
the same point of interests, such as schools, public parks, commuting areas, clubs,
commercial areas (see Fig. 1.2). The experiments confirm the presence of regular and
periodic traffic due to the tidal effect experienced by the network. In fact, it depends
not only on user habits but also by the land usage and POIs scattered in the city
area. Taking advantage of the mobile traffic patterns, we propose two mathematical
models with the goal of predicting *how* and *when* to optimize the resource allocation
in the underlying physical network. The methodology and the detailed results can be
found in Chaps. 4 and 5 of the Ph.D thesis and in the following papers: [15–17].

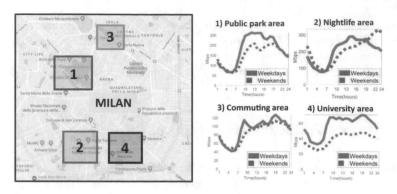

Fig. 1.2 Typical mobile traffic patterns in the city of Milan [15]

1.4 Reinforcement Learning for Adaptive Network Resource Allocation

While in the previous sections we investigated machine learning algorithms for network optimization based on supervised and unsupervised learning, in this section, we investigate the application of RL for performing dynamic Service Function Chain (SFC) resources allocation in SDN/NFV enabled metro-core optical networks. RL allows to build a self-learning system able to solve highly complex problems by employing RL agents to learn policies from an evolving network environment. Specifically, we build an RL system able to optimize the resources allocation of SFCs in a multi-layer network (packet over flexi-grid optical layer). An SFC is composed of an ordered sequence of Virtual Network Functions (VNFs). Therefore, given a set of SFC provisioning requests and their QoS requirements, the proposed method finds the optimal positioning of the VNFs in the service layer and allocates the network resources of the IP/MPLS and optical layer, respectively. Furthermore, in order to meet the SFC traffic variation over time, it dynamically predicts if and when to reconfigure the SFCs that no longer meet the required QoS criteria. Each reconfiguration implies an availability penalty on the service as it requires a temporary service interruption to allow for VNF migration and/or network resource reallocation. Thus, the mechanism presented here can be seen as a way of spending reconfiguration tokens over the span of a day. The proposed RL-based algorithm aims to spend these tokens in order to minimize the blocking probability of traffic requests finding the best action in the trade-off between availability and performance. Reinforcement learning is an approach to machine learning that trains algorithms using a system of positive and negative rewards. An RL algorithm, or agent, learns how to perform a task by interacting with its environment. The interaction is defined in terms of specific actions, observations and rewards. As we can see from Fig. 1.3, an agent performs an action at time t based on the reward and state (or observations) obtained by the environment. The action performed by the

Fig. 1.3 Schema of the proposed RL system [18]

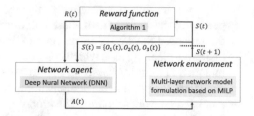

agent will produce another state of the environment and a reward at a time t + 1 and so on.

The environment, namely network environment, consists of a multi-layer model formulation based on Mixed ILP (MILP) that, given a set of SFC requests, finds the optimal VNF placement and Routing and Wavelength Assignment (RWA). The objective function of the MILP aims to maximize the number of successfully routed SFCs, minimizing: reconfiguration penalty, blocking probability and power consumption of network elements. At each time step t, i.e. hour, the network environment exposes its state S(t) made by 3 observations: $O_1(t)$: total number of reconfigurations of each SFC, at time t; $O_2(t)$: total number of blocked requests of each SFC, at time t; $O_3(t)$: traffic volume request in Gbps of each SFC, at time t. As shown in Fig. 1.3, the agent, namely Network Agent, is a Dense Neural Network (DNN) that takes as input the state of the environment S(t) and the result of the reward function R(t). As output, the agent decides if one or more SFCs need to be reconfigured. This decision is performed by means of an action A(t) executed on the environment. The reward function represents the effect of the action performed by the agent. It depends on the number of blocked service-chain requests; R(t) is positive if the agent managed to reconfigure the network in order not to have blocked requests, otherwise it is negative. Results show that, initially the system assigns negative rewards that decreases hour by hour when the algorithm does not block the requests. After few days of training, the RL agent figured out how to maximize the reward function and keep the objective function of the MILP stable to a minimum value respect to the case in which we do not use RL. In other words, it learnt how and when to reconfigure the SFCs in order to route the traffic requests by decreasing the blocking probability. The methodology and the detailed results can be found in Chap. 6 of the Ph.D. thesis and in the following paper: [18].

1.5 Implementation of Machine Learning in Real SDN/NFV Testbeds

While in the previous sections we have proposed novel algorithms in simulated networks, in this section we introduce briefly three direct applications of ML-based algorithms on different emulated and real SDN/NFV testbed scenarios.

1. First, we developed the network planner module under the Metro-Haul European project[2] in which ML-based algorithms interact with the control plane of the proposed SDN/NFV infrastructure. We introduced the software components inside the planning tool, which compose a framework that enables intelligent optimization algorithms based on ML to assist the control plane in taking strategic decisions. We presented a demonstration of a ML-assisted routing algorithm in an emulated network scenario. The proposed framework aims at guaranteeing a fair behavior towards past, current and future requests as network resource allocation decisions are assisted with ML approaches. The detailed results can be found in Chap. 7 of the Ph.D. thesis and in the following papers: [19–22].

2. Then, we developed an experimental ML-enabled SDN/NFV service orchestrator called SENATUS, targeting research environment, as devised to support testbeds for the development and validation of network services and network planning algorithms. It implements functions for managing SDN networks as well as for deploying network services on NFV infrastructure. SENATUS introduces an interface for horizontal communication that allows interworking with experimental ML modules and optimization algorithms. In addition, we present two testbed scenarios devised for testing the SDN features of the service orchestrator. The first is a physical implementation deployed at the BONSAI laboratory of Politecnico di Milano.[3] This testbed, named Minilab, provides both SDN and NVF capabilities, taking full use of SENATUS. The second testbed is an IP-over-WDM[4] network and it has been developed in collaboration with the Scuola Superiore Sant'Anna (Pisa, Italy). The testbed is composed by an IP network over an optical network infrastructure. The detailed results can be found in Chap. 8 of the Ph.D. thesis and in the following papers: [23–25].

3. As last, we present an SD-WAN open-source solution that can be tested in real network scenarios. We have implemented active and passive monitoring methods and developed classical and RL-based algorithms to improve the network performance of enterprise networks. We present two testbeds: 1) The first one is a lab implementation of SD-WAN based on open source components such as OpenDaylight SDN controller and OpenvSwitches; 2) The second is a field deployment of such solution in a real network located in an Italian city. These testbeds target the monitoring and the traffic engineering features based on RL of an SD-WAN solution and explores different approaches to understand their advantages and limitations. Our implementations provide an overlay WAN with controlled performance in terms of delay, losses and jitter over low-cost public Internet connectivity. Detailed results can be found in Chap. 9 of the Ph.D. thesis and in the following paper: [26].

---

[2]Online: https://metro-haul.eu/.

[3]Online: https://www.bonsai.deib.polimi.it/.

[4]Wavelength Division Multiplexing (WDM).

1.6   Concluding Remarks

Machine learning is one of the game-changing technologies which only recently has become mature. With the growing availability of data, computing power and platforms, ML is now applicable within the academic world and by large companies. It is an extremely powerful technique which is able to perform intelligent tasks which were up to now usually done by humans. The networking community welcomed ML as the main mathematical tool for solving complicated optimization problems. ML can recognize, predict, advice, optimize and classify, and therefore it can support the automation of various network operations enhancing the telecom industry. This PhD research work introduces the concept of Machine-Learning defined Networking. It is an approach to network management, based on SDN and NFV, that enables self-learning algorithms to improve network performance and monitoring. We have demonstrated the effectiveness of this approach by developing different algorithms able to predict and classify network traffic in order to optimize 5G metro-core networks. ML represents an essential asset for building management tools for ever increasing intelligent networks. Research on this topic is growing at a dizzying pace and is opening up research directions for many areas in telecommunications.

Acknowledgements Special thanks go to my Ph.D. supervisor Prof. Guido Maier for his continuous support and guidance.

References

1. Cisco (2019) Cisco systems. http://www.cisco.com/
2. Cisco Visual Networking Index (2016) Global mobile data traffic forecast update, 2015–2020. http://www.cisco.com/
3. Troia S (2020) Machine-learning defined networking: applications for the 5G metro-core, PhD thesis
4. His markit (2019). https://technology.ihs.com/
5. Uppal Sanjay et al (2015) Software defined WAN for dummies. Wiley, West Sussex
6. Hawilo H et al (2014) Nfv: state of the art, challenges, and implementation in next generation mobile networks (vepc). IEEE Netw 28(6):18–26
7. Yousaf FZ et al (2017) Nfv and sdn key technology enablers for 5g networks. IEEE JSAC 35(11):2468–2478
8. Musumeci F et al (2019) An overview on application of machine learning techniques in optical networks. IEEE Commun Surv Tutor 21(2):1383–1408
9. Alvizu R et al (2017) Matheuristic with machine-learning-based prediction for software-defined mobile metro-core networks. IEEE-OSA JOCN 9(9):D19–D30
10. Troia S et al (2018) Deep learning-based traffic prediction for network optimization. In: 20th ICTON, Bucharest, Romania
11. Andreoletti D et al (2019) Network traffic prediction based on diffusion convolutional recurrent neural networks. In: INFOCOM network intelligence workshop, France, Paris
12. Alvizu R et al (2018) Machine-learning-based prediction and optimization of mobile metro-core networks. In: IEEE photonics society summer topical meeting series (SUM), Waikoloa Village, HI, USA
13. TIM (2014) Big data challenge. https://dandelion.eu/datamine/open-big-data/

14. Regione Lombardia (2013) Servizi Comunali Milano. http://www.territorio.regione.lombardia.it/
15. Troia S et al (2017) Identification of tidal-traffic patterns in metro-area mobile networks via Matrix Factorization based model. In: IEEE PerCom workshops, Kona, HI
16. Troia S et al (2019) Dynamic network slicing based on tidal traffic patterns in metro-core optical networks. In: IEEE HPSR, Xi'An, China
17. Troia S et al (2020) Dynamic programming of network slices in software-defined metro-core optical networks. In: OSN, vol 36, pp 100551
18. Troia S et al (2019) Reinforcement learning for service function chain reconfiguration in NFV-SDN metro-core optical networks. IEEE Access 7(1):1–14
19. Troia S et al (2018) Machine-learning-assisted routing in SDN-based optical networks. In: ECOC, Italy, Rome
20. Troia S et al (2019) Machine learning-assisted planning and provisioning for SDN/NFV-enabled metropolitan networks. In: EuCNC, Valencia, Spain
21. Martín I et al (2018) Is machine learning suitable for solving RWA problems in optical networks?. In: ECOC, Italy, Rome
22. Martín I et al (2019) Machine learning-based routing and wavelength assignment in software-defined optical networks. IEEE TNSM 16(3):871–883
23. Troia S et al (2018) SENATUS: an experimental SDN/NFV orchestrator. In: IEEE NFV-SDN, Verona, Italy
24. Troia S et al (2019) Dynamic virtual network function placement over a software-defined optical network. In: OFC, San Diego, USA
25. Troia S et al (2019) Portable minilab for hands-on experimentation with software defined networking. In: ConTEL, Graz, Austria
26. Troia S et al (2020) SD-WAN: an open-source implementation for enterprise networking services. In: 22th ICTON, Bari, Italy

Open Access This chapter is licensed under the terms of the Creative Commons Attribution 4.0 International License (http://creativecommons.org/licenses/by/4.0/), which permits use, sharing, adaptation, distribution and reproduction in any medium or format, as long as you give appropriate credit to the original author(s) and the source, provide a link to the Creative Commons license and indicate if changes were made.

The images or other third party material in this chapter are included in the chapter's Creative Commons license, unless indicated otherwise in a credit line to the material. If material is not included in the chapter's Creative Commons license and your intended use is not permitted by statutory regulation or exceeds the permitted use, you will need to obtain permission directly from the copyright holder.

Chapter 2
Traffic Management in Networks with Programmable Data Planes

Davide Sanvito

2.1 Software-Defined Networks (SDN)

Traditional computer networks include vertically integrated devices running both the control plane (CP) and the data plane (DP). The former computes the forwarding decisions about how to handle the traffic, while the latter effectively process it. The heterogeneity of low-level vendor-proprietary configuration interfaces, together with the proliferation of specialized devices (middleboxes, e.g.. firewalls, network address translators, load balancers and deep packet inspection boxes), made networks complex and difficult to manage. The situation was further exacerbated by the need of quickly respond to network dynamics such as failures and changes in the traffic patterns, in the network topology and in the forwarding policies.

Software-Defined Networking (SDN) is a computer networking paradigm based on the decoupling of the control plane from the data plane. The control plane is logically centralized in an external node (called the network controller) which configures the network devices through a well-defined application programming interface (API). Network devices (e.g.. switches) become simple elements running just the data plane and forwarding the traffic according to the decisions taken by the external controller. Network operators can now directly operate on the global network view offered on top of the controller without designing complex distributed protocols to achieve a desired global network-wide behaviour. SDN brings to the networking domains all the software engineering best practices such as code modularity and reusability. SDN enables the reconfiguration of the network at software speed by simply running a different application on top of the controller. The controller is in charge of maintaining the global view of the network status and can even provide to the applications running above an abstract view of the network by limiting or transforming the observable

D. Sanvito (✉)
Dipartimento di Elettronica, Informazione e Bioingegneria (DEIB),
Politecnico di Milano, Piazza Leonardo da Vinci, 32, 20133 Milan, Italy
e-mail: davide.sanvito@polimi.it

© The Author(s) 2021
A. Geraci (ed.), *Special Topics in Information Technology*,
PoliMI SpringerBriefs, https://doi.org/10.1007/978-3-030-62476-7_2

topology in order to provide isolation to different applications. At the same time the emergence of new network programming abstractions and high-level programming languages simplifies the network management and facilitates the evolution of the network. The augmented programmability brought by SDN enables a wide range of network applications ranging from traditional network functions (e.g.. switching and routing) to traffic engineering (e.g.. load balancing, QoS and fault tolerance), from network monitoring to security domains (e.g.. firewall, network access control, DoS attack mitigation).

OpenFlow [14] represents one of the most successful instances of the SDN principles. Despite the efforts towards an augmented network programmability started several years ago with Active Networks [10], the mainstream adoption of SDN principles took place only with OpenFlow. Thanks to its pragmatic compromise between the need for innovation from researchers and the need for closed platforms from the vendors, it became the de-facto standard programming interface. OpenFlow is nowadays supported by a large number of switches, both hardware and software, and SDN controllers. An OpenFlow switch exposes to the controller a pipeline of match-action tables. The *match-action* abstraction is a prominent example of programming abstractions provided by SDN. All the packets belonging to the same flow (defined by a set of packet field values, i.e. the *match*) are subject to the same treatment (i.e. the *action*, for example the forwarding or dropping of the packet or the modification of its header fields). The controller is in charge of configuring the intended network behaviour by filling the match-action tables with a set of flow entries. The match-action abstraction unifies different forwarding behaviours which can be quickly reconfigured at software speed. A device, for example, can be configured as a router if all the rules match on the IP destination address or as firewall if the rules match on Ethernet type, IP addresses and TCP ports. The controller can proactively provision the devices with flow rules and reactively add new ones whenever a packet does not match any of the available rules. The OpenFlow specification [4] defines the details of the format of the control channel messages exchanged among the control plane and the data plane and the set of header fields and packet actions a switch has to support.

Most recent advances in programmable network devices [8] take a step further and enable the network operator to program the packet parser and to define custom packet actions by combining primitives offered by the switching chip. Not only the switches' resources are more flexibly allocated and tailored to the traffic the network will deal with in terms of protocols, but the format itself of protocols (i.e. the set of header fields a switch can match on) is not fixed a priori from the chip vendor and can be customized by the network operator. The impacting factor for the success of these devices is that their increased flexibility does not come with penalties in the performance, cost and power compared to fixed-function chips. These devices are typically configured using an high-level language, such as P4 [7], and in principle a P4 program can fully describe an OpenFlow-enabled switch.

2.2 Control Plane Programmability

One of the prominent use cases enabled by SDN is the execution of Traffic Engineering (TE) algorithms on top of the controller. Traditional approaches from service providers design the routing considering the worst case traffic scenario. However, this leads to a network operating most of the time in sub-optimal conditions. SDN provides the needed flexibility to update the network more frequently, enabling online traffic optimization based on periodic traffic measurements and predictions in order to improve the network performance, reduce the operational costs and balance the utilization of network resources. The maximum achievable network reconfiguration rate is however limited by two aspects.

First of all, the changing nature of the traffic affects the optimality of the computed routing configuration. Despite the traffic pseudo-periodicity under some time scales due to ordinary daily fluctuations, an accurate traffic estimation is hard to achieve and, especially in case the traffic is significantly different from the expected scenarios, computing routing configurations which are too tied to specific traffic scenarios might lead to the congestion of the network or to unfeasible configurations. It is thus desirable to take into account some robustness considerations, by computing routing policies which are able to deal with multiple traffic scenarios under a same configuration.

The second limiting aspect is instead related to the low speed of flow programming in hardware. The transition across two network configurations is a critical procedure which might lead to broken connectivity, forwarding loops or violations of the forwarding policies. Ideally the network should be atomically updated, i.e. traffic should be forwarded either using the old configuration or the new one, and not some combination of the two. The decoupling of the control plane from the data plane makes however the set of the controller and the switches a complex asynchronous distributed system. Consistent network update mechanisms [15] are techniques able to deal with this problem by computing a set of intermediate network configurations to be sequentially scheduled to move the network from the current configuration state to a target configuration. These techniques ensure that in each intermediate configuration the consistency properties are guaranteed (i.e. the connectivity is preserved and either the old policy or the new one holds) while switch resources constraints are not violated (i.e. the additional flow rules installed must not exceed the memory available in the devices). Unluckily, these multi-steps mechanisms are not the only responsible for making the update process not atomic. A recent analysis on commercial SDN devices showed the existence of a mismatch between the switch-local control plane status reported to the centralized controller and the actual status of the data plane [13]. This discrepancy makes not completely trustworthy an information which is instead essential for the consistent update mechanisms to correctly schedule the various steps in the transition towards the final network configuration. Depending on the level of utilization of the flow tables and on the current load of the data plane, the state of the data plane might fall behind the control plane from seconds up to several minutes. This means that the time required to properly complete each update step increases and this contributes to further decrease the achievable rate of updates.

2.2.1 Traffic Engineering Framework

The first contribution of the thesis is the design of a centralized SDN Traffic Engineering framework, CRR,[1] to decide whether, when and how to reconfigure the network. Given a set of of measured Traffic Matrices (TMs) over a given period (e.g.. one day), we defined an optimization model which computes a set of routing configurations to be proactively applied during the following period. Traffic matrices are clustered in the traffic, time and routing domains and we compute, for each cluster, a routing configuration which is robust against variations within the corresponding discrete traffic subspace defined by each cluster. First of all we take into consideration the traffic domain, i.e. we look at the values of the demands over time. We also considered the time domain, i.e. we avoid clustering together TMs not adjacent in time that would require too frequent network reconfigurations. Finally, we also took into account the ultimate effect of the routing on the TMs in terms of network congestion. We indeed include the same metric commonly adopted to optimize the routing (the network congestion, in terms of Maximum Link Utilization) to also guide the clustering logic. By tuning the size of clusters (i.e. the minimum number of members) and their number, the optimization model can explore the trade-off between static Traffic Engineering schemes (which compute a single routing configuration on the whole TMs set) and dynamic Traffic Engineering schemes (which instead keep reconfiguring the network each time a new TM measurement is available) while at the same time coping with the practical constraint of limiting the number of network reconfigurations and guaranteeing a minimum holding time for each configuration. Since the transitions from a routing configurations to the next one are not instantaneous even in SDN, adjacent clusters are allowed to partially overlap close to their boundaries (i.e. near routing transitions). This means that each routing configuration will be reasonably good also for a small number of traffic scenarios expected to be handled by the configuration of adjacent clusters. As a side effect, overlaps help the potentially slow multi-step consistent network update mechanisms. The analysis of the influence of errors in traffic predictions showed an interesting trade-off between the cluster length and the prediction accuracy. If the quality of the prediction is good, we can afford to have a larger number of short clusters whose routing is more tailored to specific scenarios. As the prediction quality decreases, it is better to resort to less and larger clusters which are able to better deal with a large variety of scenarios (i.e. they are more robust to traffic uncertainty). This opens a promising research direction where the controller plays an active role in measuring and predicting the

---

[1] Clustered Robust Routing.

traffic evolution and in estimating a-posteriori the quality of the prediction in order to consequently tune the level of robustness provided by the routing configurations set. More details can be found in [16].

2.2.2 ONOS Intent Monitor and Reroute Service

We then evaluated how to integrate our Traffic Engineering framework, CRR, within a SDN platform. Open Network Operating System (ONOS) [5] is a production-ready open-source SDN network operating system built for Service Provider networks. ONOS provides high performance, scalability and availability thanks to its distributed core and proper abstractions to configure the network. Among the programming abstractions offered by ONOS, intents work at the highest level: developers can express high-level policies (i.e. "intentions") without worrying about how such behaviour is implemented in the network. For example, users can use a *Point-To-Point* intent to require the connectivity between a pair of nodes without providing any information on the path to be used. Intents can be tied to a specific traffic subset (providing a set of values for the packet header fields) and a treatment (a set of actions, for example packet header modifications, to be applied to all the packets the intent refers to). ONOS supports several types of intents and each one includes a compiler which enables the Intent Framework, the ONOS component in charge of handling intents, to translate the high-level policy described by the intent itself to the set of low-level rules to be installed in the network devices. The Intent Framework is also in charge of re-compiling the intents in case of topology changes (for example link failures) to fulfill the high-level policy transparently to the application submitting the intents.

Intents represent an interesting opportunity to integrate the CRR in ONOS because we can decouple the connectivity endpoints from the paths implementing the communication. An user can thus specify the sources and destinations of the traffic (i.e. the traffic matrix pairs) independently from when and how the CRR updates the corresponding paths in the network. The idea is to modify the low-level paths implementing the intents not only as a consequence of topology changes, but also considering changes in traffic statistics, according to the output of the CRR model. Even if the Intent Framework is designed to be extensible with additional intents and compilers, it individually compiles each intent based on its own information. The CRR aims instead at compiling together multiple intents to optimize a global network objective. In addition, integrating a computationally heavy component such as an optimization tool within the same machine running the controller can have a negative impact on ONOS's high performance requirements. We thus developed a new service, ONOS Intent Monitor and Reroute (IMR) service, which enhances the ONOS Intent Framework with an external plug&play routing logic running as an off-platform application (OPA), an application running in a separate process space with respect to the ONOS controller and communicating through REST APIs or gRPC. The IMR service, running within ONOS, is in charge of collecting from network devices the statistics related to the flow rules implementing the intents and to export them to the OPA by means of a REST API. In turns, the CRR will send back to the IMR service the routing configurations as scheduled by the CRR model.

We envision the following scenario: (1) an ONOS application submits intents to request connectivity (implicitly defining the endpoints of the TMs) and requires the IMR service to monitor their statistics. (2) the ONOS Intent Framework initially routes intents on their shortest paths while, at the same time, the IMR starts the monitoring process. (3) the CRR module, running as an OPA, collects the statistics of the intents to build the set of TMs to be fed to the CRR algorithm. After a given period (e.g.. one day), the OPA solves the CRR, schedules the activation of the routing configurations during the following period and at the same time keeps collecting the statistics to be fed to the next optimization round.

In order to prevent IMR from limiting the Intent Framework in recovering from failures, the routing paths provided by the OPA are treated as soft constraints: in case the suggested paths are not available when requested or a failure happens afterwards, the Intent Framework resorts to the standard shortest path computation for all the intents affected by the failure.

The decoupling of the application submitting the intents from the routing logic allows to re-use a same logic for different ONOS applications or to switch different routing logics for a same ONOS application. The CRR indeed represents one of the possible instances of a routing algorithm running as OPA and the IMR service implements a more general framework to interconnect any ONOS intent-based application to an external plug&play routing logic. This enables to re-use existing Traffic Engineering tools or develop new schemes based either on optimization tools or on Artificial Intelligence and Machine Learning. More details can be found in [18]. Our IMR service has been integrated in ONOS Nightingale version 1.13 as an official open-source contribution [3].

2.3 Data Plane Programmability

Despite the robust nature of each routing configuration, significant traffic deviations from the expected scenarios, such as network failures and congestion, need a proper handling to keep the network operating. Albeit the great speed in innovation and flexibility enabled by SDN, one of the main limitations introduced by its two-tier architecture is the strict decoupling of the data plane from the control plane. Network devices are indeed dummy devices unable to modify their forwarding behaviour without relying on the external controller, even if such changes depends entirely on events which are locally observable. This rigid separation of concerns prevents *a priori* the implementation of applications which require a prompt reaction, due to the intrinsic control channel latency and the processing overhead at the controller. Examples of such applications range from the security domain (e.g.. DDoS attacks mitigation) to the network resiliency domain (e.g.. detection and recovery of network failures) up to Traffic Engineering (e.g.. load balancing and congestion control). In addition, most of the operations supported within the data plane are stateless: each packet is forwarded according to the matching rule without any notion of the past history of the flow it belongs to. Applications which need to keep per-flow states (e.g.. load balancing, NAT or stateful firewall) have to rely on the external controller to support a state-dependent forwarding.

Recent research efforts tried to deal with these limitations by delegating part of the control back to network devices to enable a self-adaptation of the forwarding behaviour. In the context of the European H2020 project BEBA [1], I've contributed to the software prototyping of a stateful extension[2] to OpenFlow and to the design of some use case applications. Open Packet Processor (OPP) [6] enables stateful packet processing in the data plane with a programming abstraction based on Extended Finite State Machines (EFSM). Flows are associated with a persistent context which includes a state and few data variables. Packets are forwarded not only based on their header but also on the persistent state of the flow they belongs to. Each state determines a forwarding policy and transitions among states are triggered, directly in the fast path, according to time-based or packet-based events and conditions evaluated over the packet header and its context. The controller can configure the stateful packet processing in the data plane by providing the set of header fields which defines the flow, i.e. the entity for which an application needs to keep a state, and the architecture of EFSM, in terms of its transitions and its state-dependent forwarding policies. In turns the switch is able to autonomously instantiate at runtime per-flow instances of the state machine without involving the external controller. For those applications where the forwarding evolution depends only on local information, OPP can provide a significantly more scalable and faster solution compared to centralized approaches.

2.3.1   Network Failures

As discussed in Sect. 2.2.2, the ONOS Intent Framework is able to re-actively recover from failures transparently to the application which submits the intents. Even if those unplanned failure scenarios are not taken into account by the CRR during the optimization phase, the Intent Framework guarantees that connectivity is preserved, although in a potentially unoptimized fashion. In order to improve on this situation, it is possible to pre-compute backup path policies on top of each one of the Robust Routing configurations for different failure scenarios and proactively configure them in the network devices.

By exploiting, for example, the OpenFlow Fast-Failover mechanism it is possible to define an alternative forwarding policy to be activated by the switch itself when it detects a failure in the link associated to the current policy. This enables a more prompt reaction thanks to the avoidance of the external controller. The failure detection

---

[2]Our stateful extension to OpenFlow has been integrated as official open-source contribution [2] to *ofsoftswitch13* [11], an user-space open-source software switch widely used in the research community.

mechanism is however external to the OpenFlow specification without any guarantee on the detection delay and many existing solutions are based on the slow path. In addition, depending on the specific topology and on the computed backup policies, the alternative path for a specific failure might not be available locally to the node which detects the failure. These *remote* failure scenarios still require the intervention of the controller making challenging achieving carrier-grade recovery times.

We designed a scheme, SPIDER,[3] based on the advanced capabilites of stateful data planes, such as OPP, to offload to the data plane both the detection and the recovery of network failures even in the case of remote failures. SPIDER is inspired by Bidirectional Forwarding Detection (BFD) and MPLS Fast Reroute technologies and provides an end-to-end proactive protection to failures independent from controller reachability and with a guaranteed sub-milliseconds detection delay. SPIDER guarantees zero losses after the detection regardless the availability of the controller and for both local and remote failure scenarios. Schemes relying on the controller to activate the backup policy would instead have non-zero losses also during the recovery phase while waiting for its intervention.

Data packets are tagged with different values to select the proper forwarding behaviour (i.e. primary path or designated backup path) and, at the same time, to drive the evolution of the state machines. Tagged data packets are indeed used to implement the two mechanisms for the detection of the failures and their recovery (i.e. the activation of the alternative backup policies) with a fully configurable trade-off between the overhead and failover responsiveness. In addition, in the remote failure scenarios, tagged data packets are also used as an in-band signalling scheme able to trigger a state transition for state machines stored in a device distinct from the one detecting the failure itself. SPIDER is able to handle all the pre-planned single failure scenarios from the data plane. Multiple failures involving a same demand require instead the intervention of the external controller. More details can be found in [9].

2.3.2   Network Congestion

The second network scenario which challenges the strict decoupling of the data plane from the control plane is the network congestion. We focused here on data center networks for their unique characteristics in terms of topology and traffic. Data center networks typically present a multi-rooted tree topology such as Leaf-Spine or Fat-Tree to provide high bandwidth among servers under different racks and a high degree of resiliency. Inter-rack traffic is usually spread across a large pool of symmetric paths using Equal-Cost Multi-Path (ECMP). ECMP selects a path by computing a hash over the identifier of the flow (for example the transport-layer addresses and ports) so that all the packets of a same transport-level flow are consistently sent on the same path without creating out-of-order packets. The decision taken by ECMP is not

---

[3] Stateful Programmable Failure Detection and Recovery.

aware of the size of the flows and is agnostic to the congestion status of the paths, thus ECMP exhibits an ideal behaviour only when there is a large number of flows of comparable sizes with sufficient entropy across the headers [12]. In reality, traffic in data centers presents often a mice-elephant distribution in terms of flow sizes: there is a large number of small flows, but the largest quota of traffic, in terms of transmitted bytes, comes from a limited set of flows (the elephant flows). This is a problem for ECMP because it might happen that two or more large flows select the same downstream path, creating congestion. This situation affects both categories of flows: elephant flows do not get the bandwidth they might potentially achieve and at the same time they block smaller flows. It is important to quickly react to this condition especially to limit the impact on the mice flows which are typically latency sensitive. Elephant flows are instead more sensitive to the available bandwidth.

Once again, advanced stateful dataplanes offer interesting opportunities for the self-adapatation of the network, enabling a more scalable and prompt reaction compared to approaches relying on the external controller. We designed and implemented CEDRO,[4] an in-switch mechanism to detect and re-route large flows colliding on a same downstream path based on the stateful capabilities of OPP. By default inter-rack traffic is spread using standard ECMP. When CEDRO detects large flows experiencing congestion (i.e. using a path whose utilization is above a predefined threshold) it triggers their re-route on an alternative path by overriding the current choice of ECMP and forcing ECMP to select a new path without considering the current one in its pool. This new choice is permanently stored in the switch, just for those specific flows, thanks to a transition in their state machines. CEDRO can handle both local and remote congestion scenarios from the data plane without involving the external controller. The congestion scenario is *remote* when the switch detecting the congestion condition is not the same able to steer the traffic to an alternative path. For example in a Leaf-Spine topology this happens when two large flows coming from different Leaf switches are assigned to the same Spine switch and their traffic addresses the same Spine switch. By bouncing back tagged data packets we can realize an in-band signalling scheme to trigger a remote reaction in the Leaf switches from the Spine switch.

In this application, the ability to handle the congestion directly from the data plane is important for two reasons. The offloading of the detection improves the scalability because we avoid having the controller orchestrate the monitoring of the flow sizes and of the link utilizations. In addition, by offloading also the re-routing, CEDRO enables a quicker reaction because we avoid paying the control channel delay and processing at the controller which might constitute by themselves a non negligible quota of the lifetime of mice flows.

In summary, the Leaf-to-Leaf macroflow aggregate (i.e. the set of all the transport-layer microflows from a same Leaf switch to the other Leaf switches) is spread over the paths using ECMP and the choice of ECMP is overridden for a selected number of microflows. By paying a small penalty in the 99-th percentile and in the maximum Flow Completion Time (FCT), CEDRO enables to improve the average and 95-th

---

[4]Congested Elephant Detection and Re-routing Offloading.

percentile of the FCT compared to standard ECMP. Given the high number of mice flows, an improvement of the average metric is relevant for the latency-sensitive nature of such category of flows.

By integrating in CEDRO the in-switch failure detection capabilities of SPIDER, we can add a quick reaction to network failures which might be considered as an extreme case of congestion scenario. The resulting system would provide two levels of reaction. In case of network congestion the system re-routes to other paths just few flows from the rack-to-rack aggregate, while in case of a network failure the entire aggregate affected by the failure gets re-balanced over the remaining set of paths. More details can be found in [17].

2.4 Conclusions

In conclusion, in this thesis work we analyzed the network programmability opportunities for traffic management offered by the Software-Defined Networking paradigm at different layers. We started from the programmability of the control plane and exploited its global view to design a proactive and centralized Traffic Engineering framework to enable online traffic optimization based on periodic traffic measurements and predictions and showed how to integrate it in a production-ready SDN platform. In order to handle the unexpected scenarios which challenge the strict decoupling of the control plane from the data plane we designed and implemented two applications based on stateful extensions to OpenFlow. These applications complement the centralized and proactive approach based on the global state of the network with reactive distributed logic based on a partial local view of the network state and enabling a more prompt and scalable reaction compared to approaches based on the centralized control plane.

References

1. BEBA project. http://www.beba-project.eu/
2. ofsoftswitch13 official contribution GitHub repository. https://github.com/CPqD/ofsoftswitch13/tree/BEBA-EU
3. ONOS Wiki: IMR - Intent Monitor and Reroute service. https://wiki.onosproject.org/x/hoQgAQ
4. OpenFlow 1.5 specification. https://www.opennetworking.org/wp-content/uploads/2014/10/openflow-switch-v1.5.1.pdf
5. Berde P, Gerola M, Hart J, Higuchi Y, Kobayashi M, Koide T, Lantz B, O'Connor B, Radoslavov P, Snow W et al (2014) Onos: towards an open, distributed sdn os. In: Proceedings of the third workshop on Hot topics in software defined networking
6. Bianchi G, Bonola M, Pontarelli S, Sanvito D, Capone A, Cascone C (2016) Open packet processor: a programmable architecture for wire speed platform-independent stateful in-network processing. CoRR arXiv:abs/1605.01977

7. Bosshart P, Daly D, Gibb G, Izzard M, McKeown N, Rexford J, Schlesinger C, Talayco D, Vahdat A, Varghese G et al (2014) P4: Programming protocol-independent packet processors. ACM SIGCOMM Comput Commun Rev 44(3):87–95

8. Bosshart P, Gibb G, Kim H-S, Varghese G, McKeown N, Izzard M, Mujica F, Horowitz M (2013) Forwarding metamorphosis: Fast programmable match-action processing in hardware for sdn. ACM SIGCOMM Comput Commun Rev 43(4):99–110

9. Cascone C, Sanvito D, Pollini L, Capone A, Sanso B (2017) Fast failure detection and recovery in sdn with stateful data plane. Int J Netw Manag 27(2):e1957

10. Feamster N, Rexford J, Zegura E (2013) The road to sdn. Queue 11(12):20–40

11. Fernandes EL, Rojas E, Alvarez-Horcajo J, Kis ZL, Sanvito D, Bonelli N, Cascone C, Rothenberg CE (2020) The road to bofuss: the basic openflow userspace software switch. J Netw Comput Appl, 102685

12. Kabbani A, Vamanan B, Hasan J, Duchene F (2014) Flowbender: flow-level adaptive routing for improved latency and throughput in datacenter networks. In: Proceedings of the 10th ACM international on conference on emerging Networking Experiments and Technologies, pp 149–160

13. Kuźniar M, Perešíni P, Kostić D, Canini M (2018) Methodology, measurement and analysis of flow table update characteristics in hardware openflow switches. Comput Netw 136:22–36

14. McKeown N, Anderson T, Balakrishnan H, Parulkar G, Peterson L, Rexford J, Shenker S, Turner J (2008) Openflow: enabling innovation in campus networks. ACM SIGCOMM Comput Commun Rev 38(2):69–74

15. Reitblatt M, Foster N, Rexford J, Schlesinger C, Walker D (2012) Abstractions for network update. ACM SIGCOMM Computer Communication Review 42(4):323–334

16. Sanvito D, Filippini I, Capone A, Stefano P, Jérémie L (2019) Clustered robust routing for traffic engineering in software-defined networks. Elsevier Comput. Commun. 144:175–187

17. Sanvito D, Marchini A, Filippini I, Capone A (2020) Cedro: an in-switch elephant flows rescheduling scheme for data-centers. In: 2020 6th IEEE conference on network softwarization and workshops (NetSoft). IEEE

18. Sanvito D, Moro D, Gullì M, Filippini I, Capone A, Campanella A (2018) Onos intent monitor and reroute service: enabling plug&play routing logic. In: 2018 4th IEEE conference on network softwarization and workshops (NetSoft). IEEE, pp 272–276

Open Access This chapter is licensed under the terms of the Creative Commons Attribution 4.0 International License (http://creativecommons.org/licenses/by/4.0/), which permits use, sharing, adaptation, distribution and reproduction in any medium or format, as long as you give appropriate credit to the original author(s) and the source, provide a link to the Creative Commons license and indicate if changes were made.

The images or other third party material in this chapter are included in the chapter's Creative Commons license, unless indicated otherwise in a credit line to the material. If material is not included in the chapter's Creative Commons license and your intended use is not permitted by statutory regulation or exceeds the permitted use, you will need to obtain permission directly from the copyright holder.

Part II
Electronics

Chapter 3
Frequency Synthesizers Based on Fast-Locking Bang-Bang PLL for Cellular Applications

Luca Bertulessi

3.1 Introduction

Digital PLL architectures are gaining importance in the frequency synthesizer field, thanks to their versatility and scalability properties. Figure 3.1 shows a simplified scheme of a digital intensive fractional-N frequency synthesizer, also called a digital phase locked loop (DPLL). The frequency of the DCO output signal f_v is reduced by the divider in the feedback chain, generating a lower frequency signal f_d. The phase detector (PD) compares the f_d signal phase with the phase of the signal generated by the reference f_r and creates the error signal e that through the digital filter controls the DCO frequency tuning word tw.

The frequency control word (FCW) of the fractional divider is quantized by a sigma delta modulator before being provided to the multi-modulus divider. To avoid spurs in the output spectrum the quantization noise introduced in the loop by this operation is cancelled-out by a digital to time converter (DTC), once the DTC transfer function gain is properly matched by digital correction algorithms [1]. Digital algorithms, in fact, provide the possibility to correct analog mismatches, non-linear transfer functions and PVT variations with small headroom in terms of area and power consumption with respect to fully analog systems. Without the quantization noise and considering a Type-II loop system the phase error between f_r and f_d is dominated by random noise. This property allows implementing a small dynamic range phase detector such as the 1-bit phase detector that reduces the power consumption and the area occupation with respect to the multi-bit phase detector approach or the standard charge-pump based PLL. The adoption of a DTC combined with the 1-bit phase detector, called Bang-Bang Phase Detector (BB-PD), and digital corrections enable then low power and low spur fractional-N frequency synthesis [1]. One of the main challenge in the design a low power sub-6GHz Bang-Bang Digital PLL is to reduce

L. Bertulessi (✉)
Politecnico di Milano, Piazza Leonardo da Vinci 3, 20133 Milan, Italy
e-mail: luca.bertulessi@polimi.it

© The Author(s) 2021

A. Geraci (ed.), *Special Topics in Information Technology*,
PoliMI SpringerBriefs, https://doi.org/10.1007/978-3-030-62476-7_3

the locking time and meet at the same time the stringent phase noise mask. The locking time is exacerbated by the limited dynamic range of the Bang-Bang phase detector and mainly depend from the loop filter parameter.

3.2 Digital PLL: Output Phase Noise and Locking Transient

In a digital locked loop as the one in Fig. 3.1 the digital filter and algorithms are clocked by a reference signal f_r and the system should be analyzed as a discrete time quantized system. A fair representation of this system can be done using a multi-rate discrete time model. But neglecting the folding effect and converting the multi-rate model at the DCO sampling rate, it is possible to derive the simplified model depicted in Fig. 3.2 that can be used to compute the noise transfer function of the reference phase noise Φ_r and DCO phase noise Φ_v to the output phase noise Φ_{out} [2].

In the represented model T_R is the reference period, K_{BPD} is the gain of the phase detector, α and β are the loop filter parameters, K_T is the DCO period gain and N is the division factor. For a small PLL bandwidth, the equivalent continuous time $G_{loop}(s)$ gain of the model is

$$G_{loop}(s) \approx K_{BPD} \left(\beta + \frac{\alpha}{sT_R} \right) \frac{K_T N}{sT_R} . \tag{3.1}$$

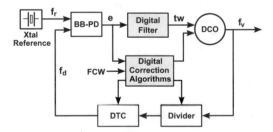

Fig. 3.1 Generic architecture of a digital phase locked loop (DPLL), synthesized standard-cell-based digital blocks are depicted in grey color

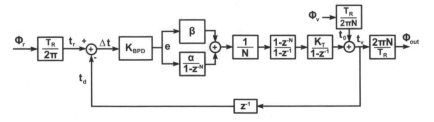

Fig. 3.2 Equivalent model of an digital phase locked loop (DPLL) at the DCO sampling rate

The main assumption in a Bang-Bang PLL design procedure is that in the steady state condition the 1-bit phase detector is working in the random noise regime. In this condition the phase error Δt is close to zero and the gaussian distributed random noise toggles the BB output between $+1$ and -1. Under this assumption and with $\sigma_{\Delta t} \gg N\beta K_T$ the K_{BPD} value inside the $G_{loop}(s)$ is related to the variance of the Δt signal [2] by the relation:

$$K_{BPD} \approx \sqrt{\frac{2}{\pi}} \frac{1}{\sigma_{\Delta t}} \ . \tag{3.2}$$

Due this dependence between the $G_{loop}(s)$ and the variance of the Δt, the mathematical optimization of the output jitter is more complex with respect to the fully analog PLL. In fact, the jitter $\sigma_{\Delta t}^2$ of the Δt signal is composed of the jitter from the reference signal $\sigma_{t_r}^2$ and the jitter from the feedback path $\sigma_{t_v}^2$.

By changing the bandwidth the output jitter also changes and so does the noise level at the phase detector input. This variation affects the $G_{loop}(s)$ and therefore the PLL bandwidth. Moreover, the value of $\sigma_{\Delta t}^2$ is the sum of the random noise jitter $\sigma_{\Delta t,rn}^2$ plus the jitter related to the limit cycle $\sigma_{\Delta t,lc}^2$, that is the periodic behavior of the state variable Δt induced by the loop quantization. The jitter due to the limit cycle is relevant only in the case of low phase noise. Considering no latency in the loop the $N\beta k_T$ quantization sets the minimum achievable input jitter

$$\sigma_{\Delta t,lc} \simeq \frac{N\beta K_T}{\sqrt{3}} \ . \tag{3.3}$$

As explained in [3] the optimum output jitter can be found when $\sigma_{\Delta t,lc} \lesssim \sigma_{\Delta t,rn}$ Choosing $\beta K_T N$ to keep the system in a random noise regime with low noise requirements, has an effect also on the locking time. To properly understand the transient behavior we have to further simplify the model in Fig. 3.2, taking into account also the non-linear characteristic of the phase detector. In fact, during the locking transient, the Δt saturates the phase detector and the random noise condition is not more valid. The phase detector can be represented with a sign function and the locking time can be evaluated considering the transient of the DCO period $T_{DCO}[k]$ that is composed by the free running period T_0 and the tuning component $tw[k] \, K_T$. The equivalent model is depicted in Fig. 3.3. When the divider modulus changes from N to N+1 the additional cycle accumulated in the feedback counter increase Δt error

Fig. 3.3 Generic DCO sampling rate model of an digital phase locked loop (DPLL) for transient analysis

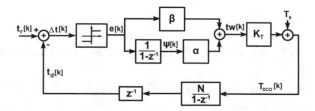

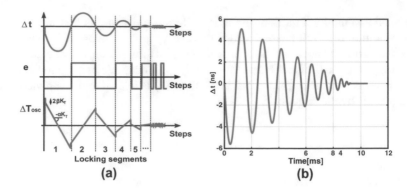

Fig. 3.4 Locking transient: (**a**) behavior of the main loop state variables, (**b**) Δt long locking transient for a noise optimized PLL

saturating the phase detector. The constant output error is integrated by the loop filter and the frequency of the oscillator is changed to reduce the time error. Each time that the sign of the error e changes, the transient enters in a new locking segment and the absolute value of Δt is reduced. When the absolute value of the time error is comparable with the time error in the steady state condition, the system reaches the locking state. The overall locking transient behavior, composed by the the locking segments, is depicted in Fig. 3.4a.

Defining the deviation of the DCO period from the final steady state DCO period $T_{DCO}[\infty]$ at each discrete step k as $\Delta T_0[k]$ it is possible to demonstrate that the locking time is proportional to:

$$T_{locking} \propto T_R \frac{1}{R(2-R)} \left(\frac{\Delta T_0}{\beta K_T} \right)^2 \tag{3.4}$$

Where R is the ratio between α and β. Comparing random noise condition ($\sigma_{\Delta t,lc} \lesssim \sigma_{\Delta t,rn}$) to the locking time (3.4), we can easily see that reducing βK_T improves the minimum jitter achievable by the system but heavily affects the locking time. With the PLL parameters carefully chosen to optimize the noise, the locking transient takes an unacceptable amount of time. For example, the locking time estimated with (3.4) for a frequency step of 100 kHz is around 9.3 ms as in Fig. 3.4b.

One way to overcome this trade-off is to change the architecture by having two separate loops: one designed for the steady-state random noise condition and the other for speeding-up the locking transient.

3.3 Multi-loop Architecture for Fast Locking Transient

A frequency synthesizer used as a local oscillator usually has to cover a wide tuning range to properly downconvert different standards or channel to the baseband. For example a 3.7 GHz DCO with a 10% of tuning range should cover a frequency range from 3.5 to 3.9 GHz. To cover this range with a single analog controlled varactor, keeping at the same time a fine frequency resolution (e.g. 10 kHz/LSB) is not feasible. In fact, to build the entire oscillator based on a single bank of minimum size digitally switched capacitive cells, leads to thousands of control wires and connections, spoiling the DCO performance in terms of area and noise. The commonly used approach is to design a segmented DCO, the entire tuning range is splitted into more overlapped tuning segments.

The fine DCO thermometric capacitive bank covers the tuning characteristic of one segment, while a coarser thermometric capacitive bank shifts this tuning segment up or down to create the overall DCO tuning characteristic. The number of elements for each segment and the relative gain K_T are limited by performance considerations and to avoid having a blind frequency region in the DCO tuning characteristic. A common rule of thumb is to size the LSB of each bank equal to half of the dynamic range of the immediately smaller bank. This ensures an overlap between two adjacent segments of more than the 50% and binds the maximum value of K_{T1} to the K_T value. The segmentation, properly controlled by a digital counter, implements different K_T gains that could be exploited to speed-up the locking transient.

From (3.4) we know that the greater the initial DCO period error, the longer the locking transient is. Looking at the equation that describe the the phase error trajectory $\Delta t[k] = t_r[k] - t_d[k]$

$$\Delta t[k] = k(T_{ref} - T_0) \pm \beta K_T N - \sum_{i=0}^{k-1}(\alpha(\Psi_0 \pm i) \pm \beta)K_T N + T_0 N$$

we also know that the maximum time error in a locking transient is non-linear with respect to the initial DCO period error.

To detect the situation of a long locking transient we can insert an additional phase detector that indicates when the phase error is above a defined threshold. The idea is to use a phase detector with a dead zone. During the steady state the time error is inside the dead zone and the output is zero. Outside the dead zone the phase detector behaves as the BB-PD controlling, with a digital PI filter, a coarser DCO bank. Moreover, this additional path has a larger value of $\beta_1 K_{T1}$ with respect to the main loop, thus allowing a fast transient without compromising the output phase noise. This path will be denoted as a frequency aid branch. The resulting architecture is reported in Fig. 3.5.

As a first analysis of the model, we may conclude that the $\beta_1 K_{T1}$ value does not have limits since this branch is disabled by the dead-zone in steady state condition. But in a practical implementation, as we have seen in the DCO segmentation design phase, K_T and K_{T1} are bonded together and this limits the effectiveness of this scheme. Moreover the presence of large quantization on the frequency-aid path due the discrete number of capacitor in DCO bank causes a large deviation from the standard trajectory and may lead to instability. The quantization and the finite dynamic range are taking into account by the model of Fig. 3.6 by the presence of the

Fig. 3.5 Nested loop PLL
architecture: the
frequency-aid branch is
composed by a TPD with
deadzone and a PI filter

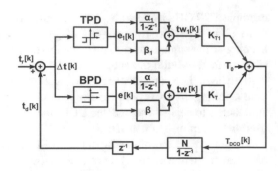

Fig. 3.6 Proposed PLL
architecture: nested loop
with feed-forward path and
deadzone

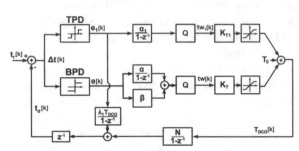

Q blocks and the limiter blocks. In order to reduce the influence of these two effects on the locking behavior of the loop and to set the gain for the frequency-aid without impacting on the DCO performance, it is possible to use an alternative control path that presents a quantization but has less effect on the locking transient.

For example, in a fractional-N PLL the divider is driven by a sigma delta modulator and the quantization error is cancelled-out by a digital to time converter (DTC) to keep the BB-PD in a random noise regime. The divider control word can handle integer and fractional divider N values with a small residual quantization in the loop. Controlling the fractional part of the divider FCW we can add or subtract a fractional part of the T_{DCO} period to the signal t_d, acting like an intrinsic integrator in the DCO path. The modified scheme is depicted in Fig. 3.6. Thanks to this feed-forward path, when the time error Δt is larger than the frequency-aid phase detector deadzone, the same time error is immediately reduced by $\lambda_1 T_{DCO}$ while the DCO frequency is adjusted by the coarser capacitive bank. If λ_1 is properly sized to keep the phase jump inside half of the deadzone, at each locking-aid activation the trajectory will restart with a Δt around zero and DCO frequency will decrease or increased at each reference cycle.

3.4 Measurement results

The PLL described in this chapter has been fully integrated in a 65nm CMOS process (see the die photo in Fig. 3.7) and occupies a core area of $0.61\,\text{mm}^2$. The measurements results were presented in [4].

Fig. 3.7 Die
microphotograph of
implemented sub-6GHz PLL

The implemented frequency synthesizer for the sub-6GHz range generates an output sinusoidal signal from 3.59 to 4.05 GHz. The reference signal is generated by an integrated reference oscillator working with an external quartz reference of 52MHz. From the die microphotograph it is possible to identify the Class-B double tail resonators DCO as the main contributor to the active area. This technology stack does not implement thick or ultra-thick metal, and the high LC quality factor needed to satisfy the output phase noise mask is obtained by using a large width main inductor. The measured output frequency can be controlled from 3.59 to 4.05 GHz, equivalent to a tuning range of 12%. The flicker corner frequency is 60 kHz.

The analog blocks are placed in the space between the DCO and the crystal oscillator (XO in Fig. 3.7). These blocks are the CMOS programmable multi modulus divider (MMD), the digital to time converter (DTC) and the Bang-Bang Phase detector (a simple Flip-Flop) and they are implemented in a similar way to [1]. The 5-level Coarse TDC is instead realized with a cascade of delay cells and Flip-Flops.

The bang-bang Phase Detector, DTC and buffers are implemented in Current Mode Logic (CML), while the divider and TDC are in standard CMOS. The power supply rails of the CML and CMOS blocks are separate to avoid disturbance coupling and each one has a dedicated and integrated decoupling capacitive bank.

The total power dissipation is 5.28 mW leading to a FoM of −247.5 dB. Figure 3.8 shows the measured phase noise. The RMS jitter (integrated from 1 kHz to 30 MHz) is 182.5 fs, while the spot phase noise at 20 MHz offset is −150.7 dBc/Hz. This noise satisfies the tight GSM specifications referred to a 900 MHz carrier of −162 dBc/Hz. The worst measured fractional spur is −50 dBc.

The locking-aid algorithm was tested by changing the divider control word and measuring the transient of the output frequency.

Figure 3.9 displays the transient response for coarse and fine frequency acquisition (a) disabling and (b) enabling the frequency aid technique. The frequency discriminator block was always enabled to guarantee at least the frequency locking. In the first row, a fine frequency step of 1 MHz is performed. Without the frequency aid branch the transient is heavily nonlinear and the target frequency is reached after 7 ms. Instead, with the two frequency aid branches active, the lock condition is achieved in just 110 μs. In the second row of Fig. 3.9, a step of 364 MHz is performed.

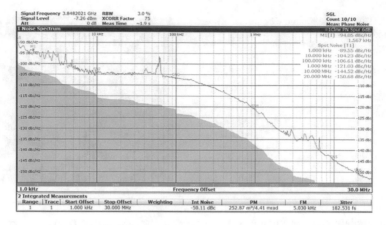

Fig. 3.8 Measured output phase noise of implemented PLL at 3.8 GHz

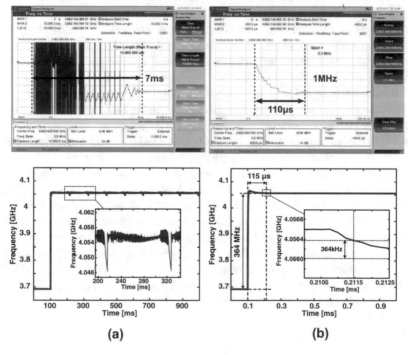

Fig. 3.9 Measurements of locking time with or without the locking-aid for large frequency steps: (**a**) without frequency aid technique, (**b**) with frequency aid technique

Disabling the frequency-aid technique (while keeping the frequency discriminator block), the circuit is unable to reach lock due to cycle slips. Enabling the frequency aid technique the PLL locks in 5.6 μs within 10 MHz from the final frequency value and takes 180 μs to fall below 1 kHz. Comparison with the state of the art is presented

Table 3.1 Comparison table with other sub-6GHz DPLL

	This work	Ref. [5]	Ref. [6]	Ref. [7]	Ref. [8]	Ref. [9]
Architecture	BB-DPLL	CPPLL	DPLL	SSPLL	DPLL	DPLL
Ref. Freq. (MHz)	52	10	26	40	40	26
Out. Freq. (GHz)	3.7–4.1	5.3–5.6	2.8–3.5	10.1-12.4	2.7–4.33	2.69
Tuning range (%)	12	6	22.2	20.4	46	n.a.
BW (MHz)	0.15	0.04	0.14	0.7	n.a.	n.a.
T_{lock} (μ s) (accuracy)	5.6/115 (10 MHz)/(364 kHz)	20 (n.a.)	n.a.	n.a.	n.a.	n.a.
Freq. Hop (MHz)	364	100	n.a.	n.a.	n.a.	n.a.
PN[a] @20MHz (dBc/Hz)	−163.3	−135.7	−162.5	−154.3	−154.0	−163.5
RMS jitter (fs)	183	n.a.	665	176	159	137
Power (mW)	5.28	19.8	15.6[b]	5.6	8.2	13.4
FoM[c] (dB)	−247.5	n.a.	−231.6	−247.6	−246.8	−246.0
Frac. Spur. (dBc)	−50	n.a.	−60	−56.6	−54	−78
Spur. Freq. (MHz)	0.5	n.a.	0.001	0.01	0.1	0.03
Area (mm^2)	0.61	1.61	0.34	0.77	n.a.	0.257
CMOS process (nm)	65	180	65	28	28	14

[a]Scaled to 900 MHz
[b]FoM $= 10\log(\sigma/1s)^2(P_{mW}/1mW)$
[c] Ref. and DCO buffers included

in Table 3.1. The table includes only the published works at the time of the ISSCC submission of [4].

3.5 Conclusions

This chapter presented the design and the optimization of both integrated output jitter and locking time in digital frequency synthesizers. Due to the limited and non-linear characteristics of the BB phase detector, the locking transient time is a common issue in the digital BB-PD PLL architecture. The proposed new locking aid techniques are able to break the trade-off between loop bandwidth and locking time. With this scheme the BB digital frequency synthesizer is able to lock in 110μs for a 1 MHz frequency step and in 115 μs for a 364 MHz frequency step without adding any look-up table or state machine at the system. The steady state jitter of the 3.7 GHz output signal is 182.5 fs and it is obtained by implementing an high-efficiency class B oscillator with double tail resonator. The overall power consumption of 5.28 mW from 1.2 V power supply leads to a power-jitter FoM of −247.5 dB.

Acknowledgements The authors would like to acknowledge Prof. Salvatore Levantino, Prof. Carlo Samori, Prof. A. L. Lacaita, Dr. Luigi Grimaldi and Dr. Dmytro Cherniak for useful discussions and design support.

References

1. Tasca D, Zanuso M, Marzin G, Levantino S, Samori C, Lacaita AL (2011) A 2.9-4.0-GHz Fractional-N digital PLL With Bang-Bang phase detector and 560 fs_{rms} integrated jitter at 4.5-mW Power. IEEE J Solid-State Circuits 46(12):2745–2758
2. Da Dalt N (2007) Theory and implementation of digital Bang-Bang frequency synthesizers for high speed serial data communications. PhD thesis
3. Marucci G, Levantino S, Maffezzoni P, Samori C (2014) Analysis and design of Low-Jitter digital Bang-Bang phase-locked loops. IEEE Trans Circuits Syst I: Regular Papers 61(1):26–36
4. Bertulessi L, Grimaldi L, Cherniak D, Samori C, Levantino S (2018) A low-phase-noise digital bang-bang PLL with fast lock over a wide lock range. In: IEEE International Solid - State Circuits Conference - (ISSCC). San Francisco, CA, pp 252–254
5. Chiu W, Huang Y, Lin T (2010) A dynamic phase error compensation technique for fast-locking phase-locked loops. IEEE J Solid-State Circuits 45(6):1137–1149
6. Weltin-Wu C, Zhao G, Galton I (2015) A 3.5 GHz digital fractional- PLL frequency synthesizer based on ring oscillator frequency-to-digital conversion. IEEE J Solid-State Circuits 50(12):2988–3002
7. Markulic N et al (2016) 9.7 a self-calibrated 10MB/s phase modulator with -37.4dB EVM based on a 10.1-to-12.4GHz, -246.6dB-FoM, fractional-n subsampling PLL. In: 2016 IEEE international solid-state circuits conference (ISSCC), San Francisco, CA, pp 176–177
8. Gao X et al (2016) 9.6 A 2.7-to-4.3GHz, 0.16psrms-jitter, -246.8dB-FOM, digital fractional-N sampling PLL in 28nm CMOS. In: 2016 IEEE international solid-state circuits conference (ISSCC), San Francisco, CA, pp 174–175
9. Yao C et al (2017) A 14-nm 0.14-psrms Fractional-N Digital PLL With a 0.2-ps Resolution ADC-assisted coarse/fine-conversion chopping TDC and TDC nonlinearity calibration. IEEE J Solid-State Circuits 52(12):3446–3457

Open Access This chapter is licensed under the terms of the Creative Commons Attribution 4.0 International License (http://creativecommons.org/licenses/by/4.0/), which permits use, sharing, adaptation, distribution and reproduction in any medium or format, as long as you give appropriate credit to the original author(s) and the source, provide a link to the Creative Commons license and indicate if changes were made.

The images or other third party material in this chapter are included in the chapter's Creative Commons license, unless indicated otherwise in a credit line to the material. If material is not included in the chapter's Creative Commons license and your intended use is not permitted by statutory regulation or exceeds the permitted use, you will need to obtain permission directly from the copyright holder.

Chapter 4
Inductorless Frequency Synthesizers for Low-Cost Wireless

Alessio Santiccioli

4.1 Introduction

The roaring demand for wireless connectivity at a low price point has, in recent years, spurred the interest for highly-integrated transceiver solutions that are able to cut down on expensive silicon area requirements. In this context, one of the major limiting factors is generally represented by the frequency synthesizer used to generate the local oscillator signal for the transceiver. Conventionally implemented as phase-locked loops (PLLs) based around LC-oscillators, they require large amounts of area due to the use of integrated inductors. Ring-oscillator (RO) based frequency synthesizers, on the other hand, ensure reduced area occupation, provide an inherent immunity to magnetic pulling and are better suited to scaling. However, they also suffer from a worse power vs. phase noise tradeoff with respect to their LC-based counterparts [1], which—leading to an undesirable degradation in the overall transceiver efficiency—prevents a more widespread adoption.

An effective way to overcome this issue, for applications that require very low integrated jitter levels but are not constrained by tight spot-noise requirements (e.g. IEEE 802.11b and high-performance clocking), is to perform an aggressive high-pass filtering of the RO phase noise. In principle, this could be achieved by increasing the bandwidth of the phase-locked loop (PLL) controlling the RO. In practice, however, the PLL bandwidth cannot be increased indefinitely, as it must remain well below the reference frequency to ensure stability [2]. The achievable level of filtering is therefore generally rather limited.

To increase the ring-oscillator phase noise filtering bandwidth beyond the limits set by conventional PLLs, two architectures have been proposed: multiplying delay-locked loops (MDLLs) [3–10] and injection-locked phase-locked loops (IL-PLLs) [11–20]. Both architectures suppress jitter accumulation by performing a periodic

A. Santiccioli (✉)
Politecnico di Milano, Piazza Leonardo da Vinci 32, 20133 Milan, Italy
e-mail: alessio.santiccioli@polimi.it

© The Author(s) 2021
A. Geraci (ed.), *Special Topics in Information Technology*,
PoliMI SpringerBriefs, https://doi.org/10.1007/978-3-030-62476-7_4

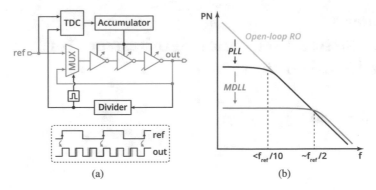

Fig. 4.1 MDLL architecture **a** block diagram and **b** phase noise filtering capabilities

realignment of the ring oscillator edges to a *cleaner* reference signal edge. Whereas in the IL-PLL case this is achieved by enforcing the crossing times of the output signal through switch transistors—which only allow for partial realignment—MDLLs rely on a multiplexer (MUX) placed within the RO loop to fully substitute a recirculating edge with the *cleaner* reference one (Fig. 4.1). Since this effectively limits jitter accumulation in the RO to only one reference cycle, MDLLs are able to achieve the highest filtering bandwidth among the two architectures, at about half of the reference frequency, $f_{ref}/2$ [21]. As a result, they clearly represent the architecture of choice for highly-efficient inductorless frequency synthesizers.

4.2 Fractional-N MDLLs

The basic MDLL architecture introduced in the previous section, inherently requires that the output frequency be an integer multiple of the input one, so that precise edge substitution can be achieved. However, to provide a viable alternative to conventional LC-based PLLs, MDLLs should also be able to generate output frequencies that are not an integer multiple of the reference one—a concept commonly referred to as fractional frequency synthesis. Unfortunately, extending the MDLL architecture to fractional-N operation presents some extra challenges.

The conventional approach to enable fractional-N frequency synthesis in MDLLs [8], is illustrated in Fig. 4.2. Similarly to a PLL, the modulus control, $MC[k]$, of the feedback divider is dithered by a $\Delta\Sigma$ modulator to achieve an average fractional division factor. For the simpler case of a first-order modulator, $MC[k]$ is switched between two levels, N and $N + 1$. This, in turn, leads to a time error between the rising reference and oscillator edges, which follows a ramp from 0 to T_v, where T_v is the oscillator period. To avoid the spectral degradation resulting from such a large quantization noise being introduced in the RO during edge-replacement, a digital-to-time converter (DTC)—which is an element that allows to introduce a

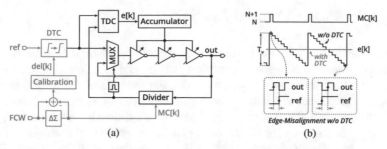

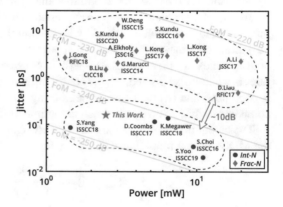

Fig. 4.2 Conventional fractional-N MDLL **a** implementation and **b** signal diagrams

Fig. 4.3 Performance gap in state-of-the-art inductorless frequency synthesizers

digitally-controlled delay on a signal—is placed on the reference path, to realign the injected edges to the recirculating RO edges. The control signal for the DTC, $del[k]$, is derived by first accumulating the $\Delta\Sigma$-quantization error to account for the intrinsic frequency-to-phase integration in the MMD, and then scaling it by a proper gain so as to match the DTC's bit-to-time conversion gain. As a result, the required DTC range is set by the amplitude of the $\Delta\Sigma$ quantization error being canceled.

Unfortunately, the DTC also degrades the reference signal by introducing both random as well as deterministic jitter, due to component noise and nonlinearities in its bit-to-time characteristic [22], respectively. Whereas in PLLs this poses a limited issue, since reference-path noise is largely suppressed by their narrow loop bandwidth, MDLLs suffer from a severe degradation in the output spectrum as a result of their much larger injection bandwidth. In fact, since the reference signal is used by MDLLs to provide a baseline for the jitter reset in the RO, DTC jitter is transferred to the output as-is. As illustrated by the plot in Fig. 4.3, this additional burden leads to a substantial performance gap between the jitter-power-product figure-of-merit (FoM) of integer-N and fractional-N inductorless frequency synthesizers, which prevents the latter from being adopted in more demanding applications.

4.3   Jitter-Power Tradeoff Analysis

To overcome the limitations of the conventional fractional-N MDLL architecture and enable low-jitter and low-power operation, it is crucial to gain an in-depth understanding for its fundamental design tradeoffs. For the case of PLLs, [25] provides appropriate guidelines to minimize the jitter-power product, in terms of an optimum loop bandwidth and power partitioning ratio among building blocks. However, since MDLLs rely mainly on edge-replacement to achieve oscillator phase noise filtering, shifting the loop bandwidth would have little effect on the overall output jitter. Therefore, an analytical expression for the jitter-power product should be first derived for the specific case of MDLL, and then analyzed to determine which degrees of freedom are available for the designer to optimize the overall system performance.

To this end, accurate yet simple expressions for the oscillator and reference path jitter contributions can be derived by leveraging the spectral estimates developed in [26] through a time-variant modeling approach. The following assumptions—which hold in almost all practical cases—will be considered for simplicity: (i) the output jitter is white-noise limited, i.e. the contribution of $1/f$ noise to the overall spectrum is negligible, and (ii) the phase noise filtering effect of the tuning loop is negligible compared to that given by the much larger injection bandwidth.

The output jitter contribution due to the RO can be derived by first approximating the MDLL output phase noise spectrum through a Lorentzian function [21]:

$$S_{\phi,ro}^{(out)}(f) = \frac{K_{inj}}{1 + \left(f/f_{inj}\right)^2} \tag{4.1}$$

where K_{inj} and f_{inj} represent the estimates derived in [26] for the low-frequency plateau and equivalent filtering bandwidth of an edge-realigned RO, given by:

$$
\begin{aligned}
K_{inj} &= \mathcal{L}_{ro}(f_{ref}) \cdot \frac{4\pi^2}{3} \frac{(N-1)(N-0.5)}{N^2} \\
f_{inj} &= f_{ref} \cdot \frac{\sqrt{1.5}}{\pi} \frac{N}{\sqrt{(N-1)(N-0.5)}}
\end{aligned}
\tag{4.2}
$$

where $\mathcal{L}_{ro}(f_{ref})$ is the single-sideband-to-carrier ratio of the free-running oscillator, evaluated at the reference frequency, f_{ref}. The corresponding output phase noise variance, $\sigma_{\phi,ro}^2$, can then be derived by symbolic integration of (4.1). Scaling the result to obtain jitter, leads to:

$$\sigma_{t,ro}^2 = \frac{\sigma_{\phi,ro}^2}{(2\pi f_{out})^2} = \mathcal{L}_{ro}(f_{ref}) \cdot \frac{1}{N f_{out} \sqrt{6}} \tag{4.3}$$

where the multiplication factor has been assumed to be $N \gg 1$. To link the jitter contribution to the respective power dissipated in the RO, P_{ro}, the commonly adopted figure-of-merit for oscillators, i.e. $\text{FoM}_{ro} = 10 \log_{10}[\mathcal{L}_{ro}(f_{ref}) \cdot (f_{ref}/f_{out})^2 \cdot (P_{ro}/1\text{mW})]$, can be substituted in the previous expression. This ultimately results in:

$$\sigma_{t,ro}^2 = \frac{10^{\frac{\text{FoM}_{ro}}{10}}}{P_{ro}} \cdot \frac{N}{f_{out}\sqrt{6}} \tag{4.4}$$

Since MDLLs rely on the reference edges to provide a baseline to which the RO edges are periodically reset [26], the output jitter contribution due to the reference path is, instead, transferred from the input as-is, i.e. $\sigma_{t,ref}^2 = \sigma_{t,in}^2$. To link also this contribution to the corresponding power consumption, an appropriate figure-of-merit can be introduced. Under the assumption that the reference path components (i.e. DTC and buffers) are CMOS-based, their jitter variance can be shown to be proportional to the reference clock frequency and inversely to the dissipated power [23]. This suggests the following figure-of-merit:

$$\text{FoM}_{ref} = 10 \log_{10}[(\sigma_{t,ref}^2/1\text{s}^2)(1\text{Hz}/f_{ref})(P_{ref}/1\text{mW})] \tag{4.5}$$

As a result, the reference path jitter contribution can be expressed as:

$$\sigma_{t,ref}^2 = \frac{10^{\frac{\text{FoM}_{ref}}{10}}}{P_{ref}} \cdot \frac{f_{out}}{N} \tag{4.6}$$

To derive an expression for overall jitter-power product figure-of-merit (FoM) [25] for MDLLs, (4.4) and (4.6) can be summed and multiplied by the total power consumption, $P_{ro} + P_{ref}$, leading to:

$$10^{\frac{\text{FoM}}{10}} = N(1+R) \cdot \frac{10^{\frac{\text{FoM}_{ro}}{10}}}{f_{out}\sqrt{6}} + \frac{1}{N}\left(1 + \frac{1}{R}\right) \cdot f_{out} \cdot 10^{\frac{\text{FoM}_{ref}}{10}} \tag{4.7}$$

where the ratio between reference path and RO power has been defined as $R = P_{ref}/P_{ro}$. Given that the reference and RO contributions in (4.7) exhibit opposite dependencies on N and R, it is reasonable to assume that a *global* minimum for the jitter-power product may indeed exist. To determine its value, the partial derivatives of (4.7) with respect to N and R are taken and set to zero. The resulting system of two equations in two unknowns can be solved for N and R, leading to the following expressions for their optimum values:

$$\begin{cases} N_{opt} = \sqrt[4]{6} \cdot f_{out} \cdot 10^{(\text{FoM}_{ref}-\text{FoM}_{ro})/20} \\ R_{opt} = 1 \end{cases} \tag{4.8}$$

That is, the lowest jitter-power product is obtained when oscillator and reference path power dissipation are balanced, i.e. $P_{ro} = P_{ref}$, and when an optimum reference frequency (i.e. the multiplication factor, N) is selected. The corresponding expression of the optimum jitter-power-product figure-of-merit can be found by plugging (4.8) into (4.7), which results in:

$$\text{FoM}_{opt} = \frac{1}{2}\left[\text{FoM}_{ref} + \text{FoM}_{ro}\right] + 4\,\text{dB} \qquad (4.9)$$

Since the optimum FoM value in (4.9) is proportional to the sum of the individual RO and reference FoMs, the system efficiency can in principle be further improved by acting on either of those two quantities. In practice, however, the ring-oscillator component is bound by thermodynamic limits to a minimum value of $-165\,\text{dB}$ [28], which can hardly be improved. The reference path, on the other hand, contains a DTC to operate the MDLL in fractional-N mode, which provides additional degrees of freedom to be leveraged. In fact, the analysis presented in [23] suggest two key guidelines to this regard:

- CMOS DTCs should be preferred over fully-differential implementations, since their jitter-power performance is remarkably superior in typical application cases;
- For a given DTC architecture, reducing the required delay-range provides the main and most effective way to decrease jitter and thus improve FoM_{ref}.

In addition to the jitter-power product, several other DTC design-tradeoffs benefit from a reduction of its range as well. DTC nonlinearity, for example, also depends on the delay-range [29]. Reducing it has therefore a positive impact on linearity and, in turn, on calibration complexity and fractional-spur performance. Furthermore, since the individual delay-cells typically dominate the area required for a given DTC design, reducing the range is also beneficial to the area occupation.

4.4 DTC Range-Reduction Technique

As outlined in the previous section, reducing DTC range entails several advantages for fractional-N MDLL design. Nevertheless, given that proper edge-synchronization has to be preserved in order not to degrade the output spectrum, achieving any significant range-reduction represents a nontrivial task. To overcome these limitations, Fig. 4.4 introduces a technique that—by acting on both the injection path as well as the tuning loop—allows to achieve a substantial reduction in DTC range, without incurring in any edge-misalignment issues [10].

In regard to the injection path, range reduction is achieved as follows. Assuming that the oscillator duty-cycle is 50%,[1] an opposite polarity edge is available every $T_v/2$. In principle, since only an alignment to the nearest edge is necessary for the injection to be performed correctly, the DTC range can be reduced to $T_v/2$. If a specific RO edge then happens to be of opposite polarity with respect to the reference one, correct realignment can still be recovered by leveraging a differential oscillator

---

[1]This assumption will be relaxed in the next section, where an automatic correction circuit is introduced to account for non-50% duty-cycles.

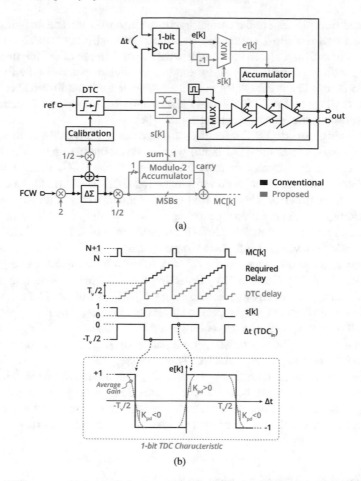

Fig. 4.4 DTC range-reduction technique **a** schematic and **b** corresponding signal diagrams

implementation, and simply swapping the injected signal around. The corresponding signal diagrams, for the simpler case of a first-order $\Delta\Sigma$-modulator, are shown in Fig. 4.4b. Conventionally, the delay required from the DTC follows a ramp from 0 to T_v, as a result of the quantization noise amplitude introduced by the $\Delta\Sigma$-modulator. By resetting the DTC control word in the second part of the delay-ramp, i.e. after a maximum $T_v/2$ delay has been reached, the rising reference edges become aligned with falling edges in the oscillator. To match edge polarity, the reference signal is then swapped around according to the value of a control signal, $s[k]$, which is set to 1 during the second part of the ramp.

The $s[k]$ control signal is derived via a successive requantization of the frequency-control word (FCW), as shown in Fig. 4.4a. A multiplication by two (i.e. a shift left) is first performed on the input FCW, so that all bits of the fractional part—except for the MSB—are requantized by the first $\Delta\Sigma$-modulator. Its output is then divided by two (i.e. shifted right) to restore the correct fractional information. The resulting signal is then fed to a modulo-2 accumulator—which essentially behaves like a single-bit first-order $\Delta\Sigma$-modulator—to complete the requantization of the fractional part, providing a dithered control signal for the integer divider placed in the frequency acquisition loop. The accumulated quantization error from the first $\Delta\Sigma$ is used as new control signal for the DTC, whereas the sum output of the modulo-2 accumulator finally represents the inversion-control signal, $s[k]$.

In the tuning loop, the DTC-reset method described so far would lead to a square-wave-like time error at the TDC input, with a corresponding amplitude of $T_v/2$, which is caused by the missing delay introduced via the DTC. To maintain lock even under these conditions, a power-hungry multi-bit TDC would generally be required to track the error. Then, to avoid spurious modulations of the RO, this error would additionally require proper canceling at the TDC output. To overcome these issues and allow for low-power and low-jitter operation, a 1-bit TDC operated in sub-sampling mode is leveraged as follows. Conventionally, 1-bit TDCs are used to detect time-errors in a narrow range around $\Delta t = 0$, for which they exhibit an equivalent linear gain, K_{pd} [24]. However, by connecting the TDC in sub-sampling mode—i.e. by allowing it to directly the oscillator signal instead of the divider output—the time error can be detected with respect to all oscillator edges, virtually increasing its range well above $\Delta t = 0$. In fact, as illustrated by the lower part of Fig. 4.4b, this results in a 1-bit TDC characteristic with a period of T_v and gain of opposite-sign every $T_v/2$. Therefore, the deterministic square-wave-like time error just shifts the operating point for noise detection in 1-bit TDC, either around the $\Delta t = 0$ region or the $\Delta t = -T_v/2$ one. Since both are able to provide an average linear gain, phase detection is not compromised. To recover the correct time-error sign, the 1-bit TDC output signal, $e[k]$, is then simply inverted according to the value of $s[k]$.

4.5 Implemented Architecture

Figure 4.5 shows the block diagram of the proposed fractional-N MDLL architecture, which has been implemented in a standard 65 nm CMOS process [10]. The system leverages the proposed DTC range-reduction technique and the results from the jitter-power tradeoff analysis, to achieve both low-jitter and low-power fine fractional-N frequency synthesis.

The MDLL is based around a five-stage pseudo-differential ring oscillator, which is tuned via current-starved NMOS transistors [8]. A simple pulser circuit, based on an AND-gate edge detector, identifies the rising edges of the DTC-delayed reference signal to be injected, ref$_\text{dtc}$, and controls the multiplexer accordingly. A swapping-MUX—i.e. a transmission-gate-based multiplexer with an embedded polarity reverser—is used to selectively swap the polarity of the differential injection signal, whenever $s[k] = 1$. Since static timing offsets between the injection and tuning paths would lead to reference spurs in the MDLL output spectrum, an automatic time offset compensation is additionally used [8].

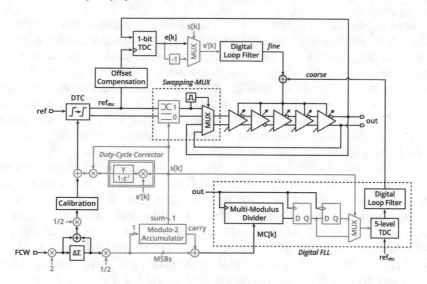

Fig. 4.5 Block diagram of the implemented MDLL prototype

Fine frequency tuning is achieved via the previously introduced 1-bit TDC sub-sampling loop. Coarse frequency acquisition is instead achieved by means of a digital frequency-locked loop (FLL), based on a variant of [27]. It relies on a low-power, five-level TDC to sense the coarse timing difference between rising edges of the DTC-delayed reference signal, ref_{dtc}, and divider output, div. The TDC output is then fed to a digital loop filter, which provides the coarse tuning information for the RO. Once locking has been achieved, the mid-thread characteristic of the five-level TDC ensures that the FLL enters an automatic dead-zone state (with negligible power consumption), which is only left if a significant phase disturbance is sensed between ref_{dtc} and div. Since the DTC range-reduction technique determines a residual $T_v/2$ time error between ref_{dtc} and div, false triggering of the 5-level TDC may become an issue in fractional-N mode. To avoid this, the $s[k]$ control signal is also used in the FLL to selectively resample the output of the integer-N divider, with either the rising or the falling edge of the oscillator (out). This effectively introduces a $T_v/2$ additional delay on the divided signal, which compensates for the reduced DTC range on the reference path.

The DTC is segmented into a coarse- and a fine-resolution stage, both of which are based on a CMOS-implementation in order to improve the overall efficiency. The coarse DTC is implemented as a cascade of buffer cells, with an embedded multiplexer that allows to set the effective length of the delay line. The fine DTC is instead implemented by digitally varying the capacitive load of a CMOS inverter, and thus its delay. Two cross-latched inverters are then additionally used to generate the required pseudo-differential DTC output. The bit-to-time conversion gain of the two DTCs is adjusted in background by a digital calibration block, which also compensates for their nonlinearity and mismatches.

Since ring-oscillators are subject to process, voltage and temperature (PVT) variations that cause their duty-cycle to vary, rising and falling edges which will not be exactly $T_v/2$ apart. This, in turn, would lead to a misalignment between reference signal and the recirculating edges, every time a polarity reversal is performed by the swapping-MUX. To avoid the resulting degradation in the output spectrum, a least-mean-square (LMS) based duty-cycle corrector (DCC) has also been implemented. The DCC operates in background and provides an output value which, summed to the DTC control word, allows to cancel the timing mismatches between reference and RO edges through the DTC itself.

To minimize the overall jitter-power product, the MDLL multiplication factor has been chosen according to (4.8), and the power budget for the RO and reference-path components has been equalized as closely as possible. Overall, the blocks running at the reference frequency dissipate 1.64 mW at 100 MHz, and introduce about 300 fs RMS jitter, leading to $FoM_{ref} = -328$ dB. The RO, instead, dissipates 860 μW and exhibits -119 dBc/Hz phase noise at an offset of 10 MHz, leading to $FoM_{ro} = -164$ dB. As a result, the optimum value for the multiplication factor is $N_{opt} = 16$, with a corresponding expected theoretical $FoM_{opt} = -242$ dB.[2]

4.6   Measurement Results

The prototype, whose die micrograph is shown in Fig. 4.6, has been implemented in a standard 65 nm CMOS process. It occupies a total core area of 0.0275 mm^2, with 0.0175 mm^2 reserved for the digital core and 0.01 mm^2 for the analog blocks (excluding the output buffer). The system is capable of fine fractional-N frequency synthesis in the 1.6-to-3.0 GHz range, with a resolution of around 190 Hz. At 1.6 GHz, the synthesizer core dissipates 2.5 mW from a 1.2 V supply.

Figure 4.7 provides the phase noise measurement in both the integer-N and fractional-N modes, as well as the free-running ring-oscillator profile, around 1.6 GHz. The corresponding RMS jitter values (integrated from 30 kHz to 30 MHz) are 334 fs and 397 fs, for the integer-N and the fractional-N case, respectively. At 1 MHz offset from the carrier, the phase noise level is -122.37 dBc/Hz in the fractional-N mode.

Table 4.1 provides a summary of the measured performances, as well as a comparison to other state-of-the-art fractional-N inductorless frequency synthesizers. In the fractional-N mode, the synthesizer reaches a jitter-power FoM of -244 dB, achieving an almost 10 dB improvement over previous state-of-the-art, and effectively bridging the gap to integer-N implementations (see previous Fig. 4.3). The corresponding bandwidth-normalized FoM_{norm}, which accounts for the limited jitter

---

[2]This estimate assumes an infinite jitter integration bandwidth. To account for the limited integration bandwidth in measurements, a correction factor of $(2/\pi) \cdot \log_{10}[\tan^{-1}(f_{meas}/f_{inj})]$ can be introduced, where f_{meas} is the upper measurement limit and f_{inj} is the injection bandwidth [10].

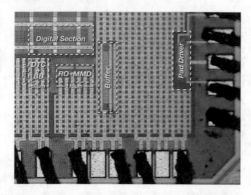

Fig. 4.6 Die micrograph of the MDLL prototype, implemented in 65 nm CMOS

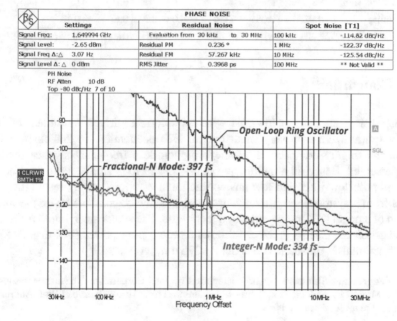

Fig. 4.7 Phase noise spectra and corresponding jitter level, measured in the fractional-N mode, integer-N mode and open-loop ring-oscillator

integration bandwidth in measurements [10], is -240 dB. The 2 dB discrepancy with respect to the theoretical -242 dB prediction derived in Sect. 4.5, is likely due to a residual power imbalance between oscillator and reference path, as well as to the Lorentzian approximation used for the spectra.

Table 4.1 Performance comparison

	This Work	Kundu ISSCC16	Marucci ISSCC14	Deng ISSCC15	Li JSSC17	Liu CICC18	Gong RFIC18
Architecture	**MDLL**	MDLL	MDLL	IL-PLL	IL-PLL	IL-PLL	IL-PLL
Freq. Range (GHz)	**1.6–3.0**	0.2–1.45	1.6–1.9	0.8–1.7	N/A	0.6–1.7	1.8–2.7
Ref. Freq. (MHz)	**100**	87.5	50	380	48	100	64
Mult. Factor (N)	**16–30**	2–16	32–38	2–4	24	6–17	28–42
Output Freq. (GHz)	**1.65**	1.4175	1.65	1.5222	1.152	0.97	2.431
Ref. Spur (dBc)	**−56**	−45	−47	−63	N/A	N/A	−43.6
Frac. Spur (dBc)	**−51.5**	N/A	−47	N/A	−57	−58.8	−45.8
Power (mW)	**2.5**	8	3	3	19.8	2.5	1.33
Int. Jitter (ps)	**0.397**	2.8	1.4	3.6	1.48	1.2	1.6
FoM (dB)	**−244**	−222	−232	−224.2	−223.6	−234.4	−234.7
FoM$_{norm}$ (dB)	**−240**	−216.6	−230.2	−219.9	−223	−226.3	−228.4
CMOS Tech. (nm)	**65**	65	65	65	65	65	40
Area (mm^2)	**0.0275**	0.054	0.4	0.048	0.6	0.12	0.13

$$\mathrm{FoM} = 10\log\left[\left(\mathrm{Jitter}/_{1s}\right)^2 \cdot \left(\mathrm{Power}/_{1mW}\right)\right] \qquad \mathrm{FoM_{norm}} = \mathrm{FoM} - 10\log\left[2/_\pi \cdot \tan^{-1}\left(f_{meas}/f_{inj}\right)\right]$$

4.7 Conclusion

The increasing demand for low-cost wireless solutions, drives the pursuit of frequency synthesizers with very small overall area occupation. In this chapter, the design of a highly compact yet efficient inductorless frequency synthesizer has been presented. Based on a multiplying delay-locked loop architecture, the system achieves both low-jitter and low-power fractional-N operation, by leveraging the results from a system-level jitter-power tradeoff analysis, combined with the introduction of a novel DTC range-reduction technique. The synthesizer, implemented in a standard 65 nm CMOS process, achieves a record jitter-power FoM of −244 dB in the fractional-N mode, in a compact 0.0275 mm^2 core area.

Acknowledgements This work has been supported by Intel Corporation. The author wishes to thank Dr. Mario Mercandelli, Prof. Salvatore Levantino, Prof. Carlo Samori and Prof. Andrea L. Lacaita for the useful discussions.

References

1. Abidi AA (2006) Phase noise and jitter in CMOS ring oscillators. IEEE J Solid-State Circuits 41(8):1803–1816
2. Gardner F (1980) Charge-pump phase-lock loops. IEEE Trans Commun 28(11):1849–1858
3. Ye S, Jansson L, Galton I (2002) A multiple-crystal interface PLL with VCO realignment to reduce phase noise. IEEE J Solid-State Circuits 37(12):1795–1803

4. Gierkink SLJ (2008) Low-spur, low-phase-noise clock multiplier based on a combination of PLL and recirculating DLL with dual-pulse ring oscillator and self-correcting charge pump. IEEE J Solid-State Circuits 43(12):2967–2976

5. Elshazly A, Inti R, Young B, Hanumolu PK (2013) Clock multiplication techniques using digital multiplying delay-locked loops. IEEE J Solid-State Circuits 48(6):1416–1428

6. Yang S, Yin J, Mak P, Martins RP (2018) A 0.0056mm^2 all-digital MDLL using edge re-extraction, dual-ring VCOs and a 0.3mW block-sharing frequency tracking loop achieving 292fs rms Jitter and -249dB FOM. In: Proceedings of 2018 IEEE international solid-state circuits conference, pp 118–120

7. Park P, Park J, Park H, Cho S (2012) An all-digital clock generator using a fractionally injection-locked oscillator in 65nm CMOS. In: Proceedings of 2012 IEEE international solid-state circuits conference, pp 336–337

8. Marucci G, Fenaroli A, Marzin G, Levantino S, Samori C, Lacaita AL (2014) A 1.7GHz MDLL-based fractional-N frequency synthesizer with 1.4ps RMS integrated and 3mW power using a 1b TDC. In: Proceedings of 2014 IEEE international solid-state circuits conference, pp 360–362

9. Kundu S, Kim B, Kim CH (2016) A 0.2-to-1.45GHz subsampling fractional-N all-digital MDLL with zero-offset aperture PD-based spur cancellation and in-situ timing mismatch detection. In: Proceedings of 2016 IEEE international solid-state circuits conference, pp 326–328

10. Santiccioli A, Mercandelli M, Lacaita AL, Samori C, Levantino S (2019) A 1.6-to-3.0-GHz fractional-N MDLL with a digital-to-time converter range-reduction technique achieving 397fs jitter at 2.5-mW power. IEEE J Solid-State Circuits 54(11):3149–3160

11. Kim H, Kim Y, Kim T, Park H, Cho S (2016) A 2.4GHz 1.5mW digital MDLL using pulse-width comparator and double injection technique in 28nm CMOS. In: Proceedings of 2016 IEEE international solid-state circuits conference, pp 328–329

12. Choi S, Yoo S, Choi J (2016) A 185fs$_{rms}$-integrated-jitter and -245dB FOM PVT-robust ring-VCO-based injection-locked clock multiplier with a continuous frequency-tracking loop using a replica-delay cell and a dual-edge phase detector. In: Proceedings of 2016 IEEE international solid-state circuits conference, pp 194–195

13. Coombs D, Elkholy A, Nandwana RK, Elmallah A, Hanumolu PK (2017) A 2.5-to-5.75GHz 5mW 0.3ps$_{rms}$-jitter cascaded ring-based digital injection-locked clock multiplier in 65nm CMOS. In: Proceedings of 2017 IEEE international solid-state circuits conference, pp 152–154

14. Megawer KM, Elkholy A, Coombs D, Ahmed MG, Elmallah A, Hanumolu PK (2018) A 5GHz 370fs$_{rms}$ 6.5mW clock multiplier using a crystal-oscillator frequency quadrupler in 65nm CMOS. In: Proceedings of 2018 IEEE international solid-state circuits conference, pp 392–394

15. Yoo S, Choi S, Lee Y, Seong T, Lim Y, Choi J (2019) 30.9 A 140fs$_{rms}$-jitter and -72dBc-reference-spur ring-VCO-based injection-locked clock multiplier using a background triple-point frequency/phase/slope calibrator. In: Proceedings of 2019 IEEE international solid-state circuits conference, pp 490–492

16. Deng W, Yang D, Narayanan AT, Nakata K, Siriburanon T, Okada K, Matsuzawa A (2015) A 0.048mm^2 3mW synthesizable fractional-N PLL with a soft injection-locking technique. In: Proceedings of 2015 IEEE international solid-state circuits conference, pp 252–254

17. Deng W, Yang D, Ueno T, Siriburanon T, Kondo S, Okada K, Matsuzawa A (2015) A fully synthesizable all-digital PLL with interpolative phase coupled oscillator, current-output DAC, and fine-resolution digital varactor using gated edge injection technique. IEEE J Solid-State Circuits 50(1):68–80

18. Li A, Chao Y, Chen X, Wu L, Luong HC (2017) A spur-and-phase-noise-filtering technique for inductor-less fractional-N injection-locked PLLs. IEEE J Solid-State Circuits 52(8):2128–2140

19. Gong J, Yuming H, Ba A, Liu YH, Dijkhuis J, Traferro S, Bachmann C, Philips K, Babaie M (2018) A 1.33 mW, 1.6ps$_{rms}$-integrated-jitter, 1.8-2.7 GHz ring-oscillator-based fractional-N injection-locked DPLL for internet-of-things applications. In: Proceedings of 2018 IEEE radio frequency integrated circuits symposium, pp 44–47

20. Liu B et al. (2018) A 1.2ps-jitter fully-synthesizable fully-calibrated fractional-N injection-locked PLL using true arbitrary nonlinearity calibration technique. Proceedings of 2018 IEEE custom integrated circuits conference, pp 1–4
21. Da Dalt N (2014) An analysis of phase noise in realigned VCOs. IEEE Trans Circuits Syst II Express Briefs 61(3):143–147
22. Levantino S, Marzin G, Samori C (2014) An adaptive pre-distortion technique to mitigate the DTC nonlinearity in digital PLLs. IEEE J Solid-State Circuits 49(8):1762–1772
23. Santiccioli A, Lacaita AL, Samori C, Levantino S (2017) Power-jitter trade-off analysis in digital-to-time converters. Electron Lett 53(5):306–308
24. Da Dalt N (2008) Linearized analysis of a digital bang-bang PLL and its validity limits applied to jitter transfer and jitter generation. IEEE Trans Circuits Syst I Regul Pap 55(11):3663–3675
25. Gao X, Klumperink EAM, Geraedts PFJ, Nauta B (2009) Jitter analysis and a benchmarking figure-of-merit for phase-locked loops. IEEE Trans Circuits Syst II Express Briefs 56(2):117–121
26. Santiccioli A, Lacaita AL, Samori C, Levantino S (2019) Time-variant modeling and analysis of multiplying delay-locked loops. IEEE Trans Circuits Syst I Regul Pap 55(11):3775–3785
27. Bertulessi L, Grimaldi L, Cherniak D, Samori C, Levantino S (2018) A low-phase-noise digital bang-bang PLL with fast lock over a wide lock range. In: Proceedings of 2018 IEEE international solid-state circuits conference, pp 252–254
28. Navid R, Lee TH, Dutton RW (2005) Minimum achievable phase noise of RC oscillators. IEEE J Solid-State Circuits 40(3):630–637
29. Elkholy A, Anand T, Choi W, Elshazly A, Hanumolu PK (2015) A 3.7 mW low-noise wide-bandwidth 4.5 GHz digital fractional-N PLL using time amplifier-based TDC. IEEE J Solid-State Circuits 50(4):867–881

Open Access This chapter is licensed under the terms of the Creative Commons Attribution 4.0 International License (http://creativecommons.org/licenses/by/4.0/), which permits use, sharing, adaptation, distribution and reproduction in any medium or format, as long as you give appropriate credit to the original author(s) and the source, provide a link to the Creative Commons license and indicate if changes were made.

The images or other third party material in this chapter are included in the chapter's Creative Commons license, unless indicated otherwise in a credit line to the material. If material is not included in the chapter's Creative Commons license and your intended use is not permitted by statutory regulation or exceeds the permitted use, you will need to obtain permission directly from the copyright holder.

Chapter 5
Characterization and Modeling of Spin-Transfer Torque (STT) Magnetic Memory for Computing Applications

Roberto Carboni

5.1 Introduction

The ubiquitous widespread of mobile devices marked the beginning of the Internet of Things (IoT) era, where the information is acquired, elaborated and transmitted by billions of interconnected smart devices. IoT demanded also a paradigm shift, from a centralized model, where acquired data was simply transmitted to a central mainframe to be processed, to a distributed model, requiring real-time data elaboration right where data is collected. For example, emerging applications such as active health monitoring, drone/robot navigation and autonomous car driving, require online elaboration of massive data. In this scenario, there is an ever increasing need for faster computing and larger/faster storage available on the IoT devices themselves.

During the last 50 years information technology achieved tremendous advancements in terms of computing power. This trend was made possible by the continuous miniaturization of the metal-oxide-semiconductor field-effect transistor (MOSFET). To describe such scaling pace, in 1965, Gordon Moore observed that the number of transistors on a silicon chip doubled every 18 months [1], speculating that such trend would continue in order to sustain the economics of electronics (Moore's law). Unfortunately, this trend is starting to slow down and is currently facing severe challenges. The so-called heat wall is one of the challenges faced by present day electronics [2], and one of the main aspects hindering Moore's law. The main reason for this high power dissipation is the increased leakage power typical of scaled transistor, which is directly connected to the Boltzmann statistics-limited sub-threshold swing of $60\,mV/dec$ [3]. The growing difficulty in keeping up with Moore's law is one of the critical aspects in hindering next generation computing. Moreover, some other challenges are not related to technology, but they are rooted in the structure

R. Carboni (✉)
Dipartimento di Elettronica, Informazione e Bioingegneria (DEIB) and IU.NET, Politecnico di Milano, Piazza Leonardo da Vinci 32, 20133 Milan, Italy
e-mail: roberto.carboni@polimi.it

© The Author(s) 2021
A. Geraci (ed.), *Special Topics in Information Technology*,
PoliMI SpringerBriefs, https://doi.org/10.1007/978-3-030-62476-7_5

of computer systems. In fact, the great majority of computer systems are based on the von Neumann architecture, which is characterized by a rigid separation of logic and memory circuits requiring a continuous movement of data between them. This condition is usually referred as memory wall or von Neumann bottleneck [4].

Aiming at the mitigation of such effect, an improved storage solution was proposed by IBM with the Storage Class Memory (SCM) concept [5]. For this purpose it should have high read/write speeds, below 100 ns like DRAM, low cost per bit, high density and non-volatility like Flash memory. Emerging memory technologies are considered a prominent candidate for SCM implementation thanks to their non-volatility, low power/fast operation and better scalability [6]. Figure 5.1 shows various emerging memory concepts, including resistive switching memory (RRAM), phase-change memory (PCM), ferroelectric memory (FERAM) and spin-transfer torque magnetic memory (STT-MRAM) [7]. Table 5.1 reports a comparison of emerging memory technologies performances. They generally depend on material-based storage, which relies on the physics of the constituent active materials. Each of them is based on its peculiar transport and switching mechanisms, although sharing the two-terminal structure, where the application of suitable voltage pulses can change one or more properties of the active material. The remarkable properties of such switching materials enables various approaches to overcome the von Neumann bottleneck,

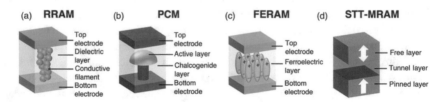

Fig. 5.1 Emerging non-volatile memory technologies. Although sharing the same two-terminal structure, the controlled modulation of the device resistance is allowed by different materials and physical phenomena. **a** In resistive-switching memory (RRAM) resistance modulation is due to the formation/disruption of a conductive filament (CF) from the top electrode to the bottom electrode. **b** In phase-change memory (PCM) amorphous-crystalline phase transition in the active layer allows for different resistance states. **c** In ferroelectric random-access memory (FERAM), the orientation of the electric dipoles in the ferroelectric layer causes a permanent polarization, resulting in different resistive states. **d** The magnetic tunnel junction (MTJ) metal-insulator-metal structure is at the core of the spin-transfer torque magnetic memory (STT-MRAM). Here, two resistive states correspond to the two relative magnetic orientation of free and pinned layer

Table 5.1 Comparative table of emerging memory technologies performances

Technology	RRAM	PCM	FERAM	STT-MRAM
Cell size[a] (F^2)	4–12	4–30	15–34	6–50
Write time (ns)	<10	≈50	≈30	1–10
Write current (μA)	10–100	80–200	<100	>50
Endurance	$>10^6$–10^{12}	$>10^9$	≈10^{10}	$>10^{15}$

[a]F represents the minimum feature size for a given microelectronics technological node

such as stochastic and neuromorphic computing, which are currently under intense scrutiny by both academia and industry [7–9].

This chapter is focused on the electrical characterization and physical modeling of STT-MRAM technology, with emphasis on its reliability and computing applications. Firstly, the dielectric breakdown-limited cycling endurance is experimentally characterized and understood with a semi-empirical model. Then, the cycle-to-cycle variability occurring in STT-MRAM is studied thanks to a physics-based model of the stochastic switching. Finally, the stochastic switching phenomenon is exploited towards the design of true-random number generator (T-RNG) and spiking neuron for stochastic/neuromorphic computing.

5.2 Spin-Transfer Torque Magnetic Memory (STT-MRAM)

Among the various emerging memory technology described in Sect. 5.1, STT-MRAM is attracting a strong interest as storage-class memory (SCM) [5, 10], DRAM replacement [11], and embedded nonvolatile memory [12], due to its fast switching, non-volatility, high endurance, CMOS compatibility and low current operation [13]. In addition, STT-RAM and spintronic devices in general can be implemented in novel non-von Neumann concepts of computing, e.g., as electronic synapse in neural networks [14], nonvolatile logic [15], and random number generator (RNG) [16].

STT magnetic memory has at its core the magnetic tunnel junction (MTJ), which consists of a metal-insulator-metal tri-layered structure comprising a thin MgO tunnel barrier ($t_{MgO} \approx 1\,nm$) separating two CoFeB ferromagnetic electrodes (FMs). One of these two electrodes, called pinned layer (PL), has a fixed magnetic polarization, whereas the free layer (FL) polarization can change between parallel (P) and anti-parallel (AP) with respect to the PL. The relative orientation of the magnetic polarization of the FL and PL determines two stable MTJ resistance states as a result of the tunnel magnetoresistance effect [13].

Specifically, the P state has a relatively low resistance R_P, while the AP state has a relatively high resistance R_{AP}. Electronic switching between the two stable resistive states takes place by spin-transfer torque (STT) effect, where the spin-polarized electrons flowing across the MTJ induce a change in the FL magnetic polarization by angular momentum conservation [17]. Perpendicular spin-transfer torque (p-STT), where the polarization of the two FMs is perpendicular to the MTJ plane, demonstrated lower switching current for the same retention time, thus enabling low-power operation and improved area scalability [18].

Figure 5.2a shows the structure of STT magnetic memory devices used for the experiments described throughout this chapter. It comprises a CoFeB PL (bottom electrode, BE) and FL (top electrode, TE) with a crystalline MgO dielectric layer. Note that the device shows an out-of-plane magnetized easy axis with two stable resistive states, namely a P-state with low resistance and an AP-state with high resistance. Figure 5.2b shows the measured current-voltage (I-V) characteristics under quasi-static ramped voltage (DC) conditions, where the set event, i.e. AP→P transition, occurred at a pos-

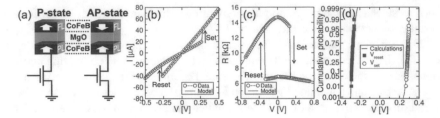

Fig. 5.2 Perpendicular STT memory device: parallel state and antiparallel state (**a**). Measured and calculated I-V curve in DC conditions (**b**) and corresponding R-V characteristics (**c**), evidencing set (AP→P) and reset (P→AP) transitions. **d** Distributions of stochastic switching voltages V_{set} and V_{reset} for 50 DC cycles. Reproduced under the terms of the CC BY 4.0 Creative Commons license from [20]. Published 2019 IEEE

itive voltage $V_{set} = 0.27$ V. The reset event, i.e. P→AP transition, occurred at a negative voltage $|V_{reset}| = 0.27$ V, underlining the symmetric switching behavior of our samples. Figure 5.2c shows the resistance-voltage (R-V) curve. The figures also show the calculated conduction characteristics by an analytical model [19].

Note that in order to drive the switching current across the MTJ, bipolar voltage pulses are applied, hence a very large electric field develops in the nanometer-thick MgO layer. As a consequence, this might induce degradation and time-dependent dielectric breakdown (TDDB) in the long term. This topic will be described in Sect. 5.3. Cycle-to-cycle repetition of the switching characteristics shows statistical variation of V_{set} and V_{reset}. Figure 5.2d shows the distributions of stochastic switching voltages V_{set} and V_{reset} for 50 DC cycles. In the thermal regime of switching, STT-induced switching takes place by random thermal fluctuations, thus featuring an intrinsically stochastic behavior. Sections 5.4–5.5 will study stochastic switching and describe some of its applications towards computing, respectively.

5.3 Understanding Dielectric Breakdown-Limited Cycling Endurance

Although the cycling endurance of STT-MRAM is sometimes referred to as virtually infinite [21], the repeated electrical stress during switching operation leads to a breakdown-limited endurance lifetime. This poses a limitation on the applicability of STT-MRAM as working memory or in-memory computing element, where extended cycling endurance is often a paramount requirement. Despite such relevant need for high endurance, the characterization methodology, the physical understanding and simulation models for breakdown-limited endurance are not yet well established. Therefore, here are presented an experimental study and a semi-empirical model of endurance failure in p-STT for the prediction of STT-MRAM lifetime.

Figure 5.3a shows the experimental set-up for the pulsed characterization of STT devices, including a waveform generator to apply triangular pulses for set (transi-

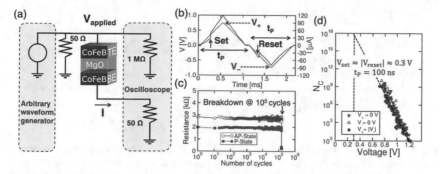

Fig. 5.3 Experimental setup for real-time monitoring of the switching characteristics during AC cycling of the p-STT devices (**a**) and measured waveforms of voltage and current (**b**). The typical voltage waveform for switching characterization and endurance experiments comprises a positive and a negative triangular pulses, which induce set and reset transitions, respectively. P and AP measured resistances during cycling (**c**), showing TMR $\approx$ 50% and endurance failure after $1.5*10^5$ cycles; median values over 10 reads are shown. After the MgO breakdown the device showed a resistance of 300 Ω, corresponding to the contact resistance. Number of cycles at endurance failure N_C as a function of the applied stress voltage for symmetric bipolar and for positive/negative unipolar cycling (**d**). Readapted under the terms of the CC BY 4.0 Creative Commons license from [22]. Published 2018 IEEE

tion from AP to P under positive voltage) and reset (transition from P to AP under negative voltage) processes, while the applied V_{TE} voltage and current I across the MTJ were monitored by an oscilloscope. Figure 5.3b shows a typical sequence of positive set, and negative reset, showing the two switching events. By monitoring the switching characteristics at each cycle, the observation of degradation phenomena and the exact event of endurance failure is possible. This event is shown in Fig. 5.3b, which reports the measured resistance during a typical pulsed experiment under symmetric switching ($V_+ = |V_-|$), as a function of the number of cycles. Data evidence clearly separate P and AP states with a TMR $= \Delta R/R_P \approx$ 50%, where $\Delta R = R_{AP} - R_P$. Cycling endurance failure is marked by an abrupt drop of read resistance, corresponding to a hard breakdown of the MgO dielectric layer, after a number N_C of cycles. Such destructive event can be explained by defect generation in MgO inducing a percolative path and thermal runaway [23]. After breakdown, the device shows a TMR of 0% and a constant resistance R $\approx$ 300 Ω which can be attributed to the metal contacts and interfaces.

Figure 5.3c shows the measured cycling endurance N_C as a function of the applied voltage with a pulse-width $t_P = 100$ ns. Three cycling conditions are compared in the figure, i.e., symmetric bipolar stress with $V_+ = |V_-|$, positive unipolar stress with $V_- = 0$ V and negative unipolar stress with $V_+ = 0$ V. N_C data for positive and negative unipolar stress show similar behaviors, evidencing a steep exponential voltage dependence with a slope $\approx$50 mV/dec for the three regimes in the figure. A simple extrapolation to the switching voltage indicates an estimated $N_C \approx 10^{18}$ at $V = 0.3$ V and $t_P = 100$ ns, which is high enough to comply with most SCM and DRAM applications. Figure 5.4a shows cycling endurance for asymmetric bipolar

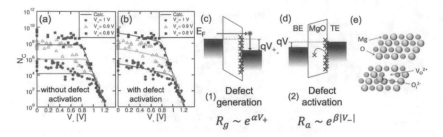

Fig. 5.4 Schematic illustration of the semi-empirical model of MgO breakdown, comprising a defect generation phase and activation. The defects could be considered to be $O_i^{2-}-V_O^{2+}$ Frenkel pairs (**e**). Measured and calculated N_C taking into account only defect generation process (**a**). **b** Calculated cycling endurance considering also defect activation process, demonstrating good agreement with experimental data. Reproduced under the terms of the CC BY 4.0 Creative Commons license from [22]. Published 2018 IEEE

stress, with variable V_- and constant $V_+ = 1$, 0.9 and 0.8 V. The voltage dependence of N_C data shows two distinct regions, namely (i) region A for $|V_-| > V_+$ where data show a steep slope $\approx$50 mV/dec, and (ii) region B for $|V_-| < V_+$ with reduced slope $\approx$600 mV/dec. In order to describe the dependence of N_C on the voltage amplitude of the applied signal, a semi-empirical model of cycling endurance was developed [19, 22]. In this model, N_C is inversely proportional to the defect concentration within the MgO layer, namely $N_C = N_{C0}(n_D/n_{D0})^{-1}$, where N_{C0} and n_{D0} are constant and n_D was calculated as $n_D = n_{D,TE} + n_{D,BE}$, where $n_{D,TE}$ and $n_{D,BE}$ are the defect concentrations originating from the TE interface and the BE interface, respectively. Defect concentrations are given by $n_{D,TE} = n_{D0}*R_{TE}/R_0$ and $n_{D,BE} = n_{D0}*R_{BE}/R_0$, where R_{TE} and R_{BE} are the generation rates describing the cycling-induced degradation at the TE and BE interfaces, respectively, while R_0 is a constant. In crystalline MgO layer, defects might be attributed, e.g., to Frenkel pairs of O vacancies V_O^{2+} and O interstitials O_i^{2-} as shown Fig. 5.4e. As depicted in Fig. 5.4c, d, tunneling electrons are considered to have a primary role in MgO degradation according to a 2-stage mechanism, including (1) defect generation (Fig. 5.4c) and (2) defect activation (Fig. 5.4d).

In this model, defects are firstly generated by bond breaking due to tunneling electrons releasing their kinetic energy E to the lattice. Here, defect generation probability is assumed to increase exponentially with the energy E, thus the generation rate is given by $R_{TE} = R_0\exp(\alpha V_+)$, where α is a constant. Similarly, the generation rate at the BE interface can be written as $R_{BE} = R_0\exp(\alpha|V_-|)$. As demonstrated in Fig. 5.4a, the model correctly describes the steep decay of N_C in region A, however the model fails to predict the weak voltage dependence in region B. To account for the impact of the smaller voltage in the MgO degradation, we considered the defect activation mechanism displayed in Fig. 5.4d. After a positive pulse of voltage V_+, the application of a negative pulse with amplitude $|V_-| < V_+$ can activate the defects generated by the positive semi-cycle, e.g. by displacing an interstitial oxygen ion away from the corresponding O vacancy in the newly formed Frenkel pair.

Figure 5.4b reports calculated cycling endurance with both defect generation and activation, indicating a better agreement with data in both regions A and B.

To complete the endurance model, defect generation and activation at both interfaces is considered. Moreover, an explicit dependence on the pulse-widths t_+ and t_- of the positive and negative pulses, respectively. The total defect density due to generation and activation is thus written as:

$$n_D = n_{D0} \left[\frac{t_+}{t_0} e^{\alpha V_+} e^{\beta |V_-|} + \frac{t_-}{t_0} e^{\alpha |V_-|} e^{\beta V_+} \right], \tag{5.1}$$

5.4 Modeling Stochastic Switching in STT-MRAM

STT-based circuits for both memory and computing applications require accurate compact model for physics-based simulations. Various STT switching models rely on numerical simulations from the Landau–Lifshitz–Gilbert (LLG) equation [24], which are typically too computing-intensive for electronic circuit simulators. Thus, simple analytical compact models are the ideal candidates for such tasks [25]. Most analytical STT switching models are limited to the thermal regime (>200 ns) and the precession regime (<1 ns), even though practical STT-RAM applications mostly work in the intermediate regime [19, 20].

To better address the cycle-to-cycle statistical variation of STT switching, the write error rate (WER) was measured, i.e., the failure rate of the switching transition according to the experimental technique shown in Fig. 5.5. Figure 5.5a shows the voltage waveform applied to characterize the WER of set transition, consisting of: (1) a negative-voltage triangular pulse at $V_- = -0.7$ V to deterministically initialize the cell in the AP state; (2) a negative-voltage square pulse for reading the cell state; (3) a positive-voltage square pulse, with amplitude V_A and duration t_P, to induce the stochastic set transition (AP$\rightarrow$P); and (4) a positive-voltage square pulse for a final reading to verify the cell state. Pulses (1), (2), and (4) have the same pulsewidth of $1\,\mu$s, while the set pulse has a variable pulsewidth t_P ranging from 40 ns to $10\,\mu$s. The final state of the cell indicates the success or failure of the transition to P state. The WER is then defined as the ratio between the number of write failures and the total number of applied cycles. Figure 5.5b shows a similar waveform for the dual experiment, namely the evaluation of reset WER. Figure 5.5c, d shows the measured WER as a function of the voltage V applied during the stochastic square pulse for (c) reset transition and (d) set transition at increasing pulsewidth t_P. The WER drops almost exponentially at increasing voltage. As t_P decreases, the transition to low WER occurs at increasing voltage and with decreasing slope. Data in Fig. 5.5c, d are summarized in Fig. 5.5e, showing the Weibull scale parameter $V_{63\%}$, defined as the voltage for WER $= 63\%$, as a function of t_P. The time-voltage relationship in Fig. 5.5e is usually explained by the thermal model depicted in Fig. 5.5f, g. Considering AP$\rightarrow$P transition, the FL polarization experiences thermal fluctuations within the AP-state well, while the current-induced spin torque causes lowering of the PMA barrier

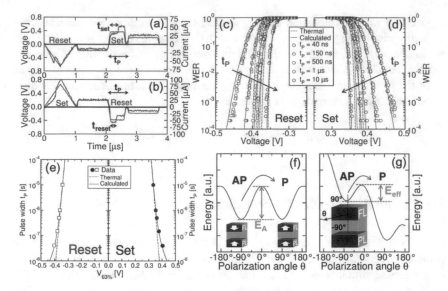

Fig. 5.5 Applied voltage waveform and corresponding read current during a WER experiment for (a) set and (b) reset transitions. **c–d** Experimental results and calculations for WER as a function of the applied voltage for different pulse widths t_P. The slope reduction for shorter t_P is correctly predicted by the new compact model. **e** Measured and calculated Weibull scale parameter $V_{63\%}$ for set and reset. PMA energy profile as a function of the FL magnetic polarization angle θ. Thermal fluctuations induce AP→P transition across the energy barrier (**f**) with no applied voltage or (**g**) with positive applied voltage. The energy unbalance between AP and P states originates from the current-induced STT. Reproduced under the terms of the CC BY 4.0 Creative Commons license from [20]. Published 2019 IEEE

E_A and the consequent transition to the P-state. Assuming a linear voltage-induced barrier lowering, the characteristic switching time is given by:

$$\tau_{th} = \tau_0 e^{\Delta\left(1-\frac{V}{V_{c0}}\right)}, \tag{5.2}$$

where $\Delta = E_A/kT$, while τ_0 and V_{c0} are constant [26]. The WER can thus be obtained by a Poissonian switching probability P given by:

$$\frac{dP}{dt} = \frac{(1-P)}{\tau}, \tag{5.3}$$

where τ is the characteristic switching time equal to τ_{th}.

Calculations with the thermal model are reported in Figs. 5.5c, d and 5.6a, b, where the Weibull plot, i.e., log(−log(WER)) is shown. Note that the thermal model correctly predicts the linear behavior for slower t_P. However, it cannot explain the deviation from the linear behavior at $t_P < 200$ ns [20, 27]. To this purpose, Fig. 5.6c, d reports the Weibull shape factor dlog(−log(WER))/dV as a function of voltage, compared with the ideal value Δ/V_{c0} (i.e., the barrier lowering coefficient) from

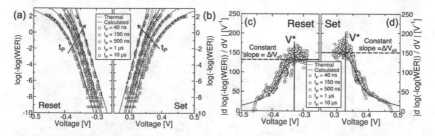

Fig. 5.6 **a–b** Voltage distributions in a Weibull plot, i.e. log(−log(WER)). Data show a deviation from the expected linear dependence in thermal regime to an anomalous non-linear shape for $t_P < 200$ ns, correctly described by the compact model. **c–d** Experimental and calculated Weibull shape factor dlog(−log(WER))/dV as a function of applied voltage V, showing a drop of the barrier lowering coefficient after a critical V*, instead of the constant value predicted by thermal model. Reproduced under the terms of the CC BY 4.0 Creative Commons license from [20]. Published 2019 IEEE

Eqs. 5.2–5.3. Data show a drop of the Weibull shape factor beyond a critical voltage $V* \approx 0.33$ V. To account for such anomalous barrier lowering at high voltage/short times, the characteristic switching time in Eq. 5.2 was rewritten as $\tau = \tau_{th} + \tau'_{th}$, where τ'_{th} is an additional time given by:

$$\tau'_{th} = \tau_0 e^{\Delta' \left(1 - erf\left(\frac{V}{V'_{c0}}\right)\right)}. \tag{5.4}$$

Figures 5.5 and 5.6 show the calculation obtained from Eqs. 5.3–5.4, demonstrating the strength of the compact model to account for WER with V-dependent Weibull shape factor in both thermal (>200 ns) and intermediate regimes (<200 ns) [20].

5.5 Stochastic STT Switching for Security and Computing

Although stochastic switching variability is harmful to the operation of STT-based magnetic memory, it is considered beneficial for emerging concept such as true-random number generator (TRNG) [16], stochastic computing [28] and brain-inspired computing [29].

On-chip generation of true random numbers is a key feature for hardware and data security for IoT. STT-based RNG can be designed as follows [16]: repeated square set/reset pulses are applied to STT cell, resulting in the stochastic set/reset events in Fig. 5.7a. Due to the stochastic t_{set}, integration of the current along the n-th set/reset cycle leads to a broadly-distributed charge $Q_n = \int I dt$ (Fig. 5.7b). The difference $Q_n - Q_{n-1}$ over two consecutive cycles (Fig. 5.7c) can have either positive or negative values with 50% probability, which are then associated to random bit values 1 or 0, respectively. Figure 5.7d, e shows a sample of generated random bits and the NIST statistical test of random bits obtained by experiments and simulation with the compact model. All tests are passed, thus supporting the feasibility of STT-based TRNG.

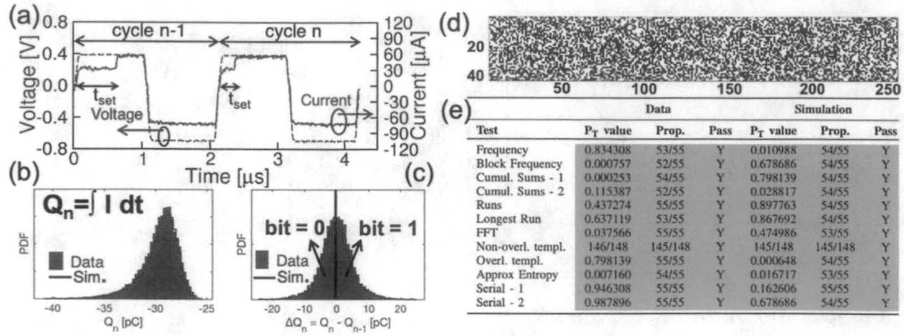

Test	Data P_T value	Data Prop.	Data Pass	Simulation P_T value	Simulation Prop.	Simulation Pass
Frequency	0.834308	53/55	Y	0.010988	54/55	Y
Block Frequency	0.000757	52/55	Y	0.678686	54/55	Y
Cumul. Sums - 1	0.000253	54/55	Y	0.798139	54/55	Y
Cumul. Sums - 2	0.115387	52/55	Y	0.028817	54/55	Y
Runs	0.437274	55/55	Y	0.897763	54/55	Y
Longest Run	0.637119	53/55	Y	0.867692	54/55	Y
FFT	0.037566	55/55	Y	0.474986	53/55	Y
Non-overl. templ.	146/148	145/148	Y	145/148	145/148	Y
Overl. templ.	0.798139	55/55	Y	0.000648	54/55	Y
Approx Entropy	0.007160	54/55	Y	0.016717	53/55	Y
Serial - 1	0.946308	55/55	Y	0.162606	55/55	Y
Serial - 2	0.987896	55/55	Y	0.678686	54/55	Y

Fig. 5.7 STT-based TRNG concept. **a** Measured square voltage pulses and current response for 2 consecutive cycles n−1 and n, (**b**) PDF of the integrated current Q_n and (**c**) PDF of differential charge $\Delta Q_n = Q_n - Q_{n-1}$. **d** Representation of 10 kb from the random bitstream(blue dot = 0, white dot = 1). **e** NIST test on 55 sequences from 1 Mb output bitstream is passed, demonstrating good randomness quality. Copyright 2019, ACM/IEEE. Reprinted with permission from [27]

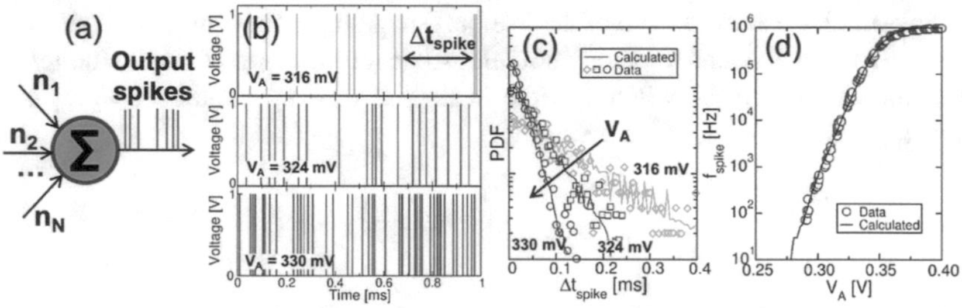

Fig. 5.8 Schematic illustration of the stochastic spiking neuron (**a**) and measured output spiking activity for increasing applied voltage V_A (**b**). Experimental and simulated exponential distributions for spike-to-spike time Δt_{spike} (**c**), demonstrating exponential spiking statistics. **d** Experimental and calculated average voltage-dependent spiking frequency Copyright 2019, ACM/IEEE. Reprinted with permission from [27]

Set/reset switching variability in STT-MRAM can also be exploited towards stochastic spiking neurons (Fig. 5.8a) [27]. In the proposed spiking neuron, a STT-MRAM cell is biased at constant voltage V_A to induce stochastic set transition from the AP state. As the switching occurs, (1) an output spike is generated and (2) the cell is reinitialized in the AP state. Figure 5.8b shows the experimental output spiking activity at increasing V_A. Spikes obey a Poissonian statistics as indicated by the exponential distributions of spike-to-spike time Δt_{spike} in Fig. 5.8c. Figure 5.8d shows the average spiking frequency $f_{spike} = 1/<\Delta t_{spike}>$ at increasing voltage, demonstrating a voltage-controlled spiking neuron.

Note that the compact model for stochastic switching (Sect. 5.4) fully supports the design and simulation of STT devices for security and computing, as demonstrated by the calculations shown in Figs. 5.7 and 5.8.

The applicability of the model to computing with spiking neurons was demonstrated for analogue multiplication (stochastic computing) and SRDP-based [30] spiking neural network [27].

5.6 Conclusions

In this chapter, STT-MRAM are described as a promising candidate for storage-class memory concept and emerging computing applications. To fully understand the applicability and limitations of this emerging memory technology, various reliability aspects were thoroughly studied. Firstly, a comprehensive study of breakdown-limited cycling endurance was presented. Cycling endurance was experimentally monitored as a function of the pulse amplitude, polarity and timing. A semi-empirical model based on defect generation and activation in the MgO tunnel barrier allowed for endurance understanding and prediction. Then, random switching in variability in STT-MRAM was characterized and described through a physics-based stochastic switching model. The compact model is capable of computing switching probability with 10^{-4} accuracy in the thermal regime (>200 ns) and intermediate regime (<200 ns), accounting for WER data of STT-MRAM as a function of the applied voltage for various pulse-widths. Finally, STT switching variability is exploited towards the implementation of true RNG, spiking neurons, analogue stochastic multiplication and SNNs. The stochastic switching compact model is demonstrated as an useful tool for design and simulation of future STT-based hardware security and computing primitives.

References

1. Moore GE (1965) Cramming more components onto integrated circuits. Electronics 38(8)
2. Ball P (2012) Computer engineering: feeling the heat. Nat News 492(7428):174
3. Salahuddin S, Ni K, Datta S (2018) The era of hyper-scaling in electronics. Nat Electron 1(8):442
4. Backus J (1978) Can programming be liberated from the von Neumann style? a functional style and its algebra of programs. Commun ACM 21(8):613–641
5. Freitas RF, Wilcke WW (2008) Storage-class memory: the next storage system technology. IBM J Res Dev 52(4/5):439
6. Burr GW et al (2008) Overview of candidate device technologies for storage-class memory. IBM J Res Dev 52(4.5):449–464
7. Ielmini D, Wong H-SP (2018) In-memory computing with resistive switching devices. Nat Electron 1(6):333
8. Carboni R, Ielmini D (2019) Stochastic memory devices for security and computing. Adv Electron Mater 1900198
9. Carboni R, Ielmini D (2020) Applications of resistive switching memory as hardware security primitive. In: Applications of emerging memory technology. Springer, Berlin, pp 93–131
10. Wong H-SP, Salahuddin S (2015) Memory leads the way to better computing. Nat Nanotechnol 10(3):191
11. Ikegami K et al (2014) Low power and high density STT-MRAM for embedded cache memory using advanced perpendicular MTJ integrations and asymmetric compensation techniques. In: IEEE international electron devices meeting
12. Shum D et al (2017) CMOS-embedded STT- MRAM arrays in 2x nm nodes for GP-MCU applications. In: Symposium on VLSI technology, pp T208–T209
13. Apalkov D, Dieny B, Slaughter J (2016) Magnetoresistive random access memory. Proc IEEE 104(10):1796–1830

14. Lequeux S et al (2016) A magnetic synapse: multilevel spin-torque memristor with perpendicular anisotropy. Sci Rep 6:31510
15. Mahmoudi H et al (2013) Implication logic gates using spin-transfer-torque-operated magnetic tunnel junctions for intrinsic logic-in-memory. Solid-State Electron 84
16. Carboni R et al (2018) Random number generation by differential read of stochastic switching in spin-transfer torque memory. IEEE Electron Device Lett 39:951–954
17. Slonczewski JC (1996) Current-driven excitation of magnetic multilayers. J Magn Magn Mater 159(1–2):L1–L7
18. Ikeda S (2010) A perpendicular-anisotropy CoFeB-MgO magnetic tunnel junction. Nat Mater 9(9):721
19. Carboni R et al (2016) Understanding cycling endurance in perpendicular spin-transfer torque (p-STT) magnetic memory. In: IEEE international electron devices meeting (IEDM)
20. Carboni R, Vernocchi E et al (2019) A physics-based compact model of stochastic switching in spin-transfer torque magnetic memory. IEEE Trans Electron Devices 60(10):4176–4182
21. Kan JJ et al (2017) A study on practically unlimited endurance of STT-MRAM. IEEE Trans Electron Devices 64(9):3639–3646
22. Carboni R et al (2018) Modeling of breakdown-limited endurance in spin-transfer torque magnetic memory under pulsed cycling regime. IEEE Trans Electron Devices 65(6):2470–2478
23. Yoshida C et al (2009) A study of dielectric breakdown mechanism in CoFeB/MgO/CoFeB magnetic tunnel junction. In: IEEE international reliability physics symposium (IRPS)
24. Xu N et al (2015) Physics-based compact modeling framework for state-of-the-art and emerging STT-MRAM technology. In: IEEE international electron devices meeting (IEDM), pp 28.5.1–28.5.4
25. Ielmini D, Milo V (2017) Physics-based modeling approaches of resistive switching devices for memory and in-memory computing applications. J Comput Electron 16
26. Heindl R et al (2011) Validity of the thermal activation model for spin-transfer torque switching in magnetic tunnel junctions. J Appl Phys 109(7)
27. Carboni R, Vernocchi E et al (2019) A compact model of stochastic switching in STT magnetic RAM for memory and computing. In: ACM/IEEE international symposium on nanoscale architectures (NANOARCH), pp 5.2.1–5.2.6,
28. Lv Y, Wang J-P (2017) A single magnetic-tunnel-junction stochastic computing unit. In: IEEE international electron devices meeting (IEDM),
29. Grollier J et al (2016) Spintronic nanodevices for bioinspired computing. Proc IEEE 104(10):2024–2039
30. Milo V et al (2018) A 4-transistors/1-resistor hybrid synapse based on resistive switching memory (RRAM) capable of spike-rate-dependent plasticity (SRDP). IEEE Trans Very Large Scale Integr (VLSI) Syst 26(12):2806–2815

Open Access This chapter is licensed under the terms of the Creative Commons Attribution 4.0 International License (http://creativecommons.org/licenses/by/4.0/), which permits use, sharing, adaptation, distribution and reproduction in any medium or format, as long as you give appropriate credit to the original author(s) and the source, provide a link to the Creative Commons license and indicate if changes were made.

The images or other third party material in this chapter are included in the chapter's Creative Commons license, unless indicated otherwise in a credit line to the material. If material is not included in the chapter's Creative Commons license and your intended use is not permitted by statutory regulation or exceeds the permitted use, you will need to obtain permission directly from the copyright holder.

Chapter 6
One Step in-Memory Solution of Inverse Algebraic Problems

Giacomo Pedretti

6.1 Introduction

Linear algebra problems, such as solving a linear system of equations, are the backbone of modern scientific computing and data-intensive tasks. Among these, machine learning is currently the discipline with most effort of study from scientists and engineers, to unleash the full power of computational algorithms with applications to any aspect of our life. These powerful algorithms, such as linear and logistic regression, are usually executed in conventional digital hardware by combining sequences of boolean functions on binary data. Thus, computing complicated operations requires a large memory and many computing steps. These problems are encoded in matrix form and executed by iteratively performing matrix-vector multiplications [7, 36], resulting in a polynomial computational time complexity, for example $O(N^3)$ where N is the size of the problem. In this chapter, novel analog circuits for the solution of matrix equations in one step will be presented. Thanks to the in-memory computing framework, the problem implementation does not require data transfer between memory and processing unit, resulting in unprecedented speed. With nanoscale crosspoint resistive memories, the novel circuit requires also less area compared to traditional technology. The results pave the way for the development of memory computing unit to overcome the limitation of current accelerators.

G. Pedretti (✉)
Dipartimento di Elettronica, Informazione e Bioingegneria, Politecnico di Milano, piazza
Leonardo da Vinci, 32, Milano, Italy
e-mail: giacomo.pedretti@polimi.it

© The Author(s) 2021
A. Geraci (ed.), *Special Topics in Information Technology*,
PoliMI SpringerBriefs, https://doi.org/10.1007/978-3-030-62476-7_6

6.2   In Memory Computing

Technology scaling has been driven by Moore's law [21] in the last decades, predicting that the number of transistors per mm^2 of an integrated circuit doubles every 18 months. Figure 6.1a depicts in red the exponential growth fitted from real data of different technology node of the last 25 years and the confirmed prediction from the future releases [28]. It is already possible to see a deviation from the ideal exponential scaling, as the some technologies will see the market with significant delay. Moore's law is in fact slowing down, due to physical limits of devices scaling and increased cost manufacturing [28]. It has also been observed [13] that the energy dissipated by transistors has decreased exponentially with the technology node until the late 80's. However, by now this energy should have reached the thermal fluctuation kT, known as Landauer limit [13], which is impossible with modern Complementary-Metal-Oxide-Semiconductor (CMOS) technologies. Even future predictions are far away from the Landauer limit. It is thus evident that new computing technologies need to be developed to keep the pace with Moore's Law and reduce the energy dissipation. Among this, novel memory devices have attracted research interest also from the computing community [39], in fact they have been demonstrated able of performing traditional computing tasks such as boolean function [31].

However Moore's law speed is not enough. Figure 6.1a shows a comparison of Moore's law (in red) with the performance required for executing state of the art algorithms (in blue) developed across the last years [3]. It is possible to see that with an exponential growth of the number of floating point operations per second (FLOP/s) that doubles every 3.4 months, the required resources scaling outperforms Moore's Law. This suggest that not only new materials and devices are needed to fulfill Moore's law requirements, but a shift of paradigm in architecture is needed to outperform traditional computing systems. Figure 6.1b shows the conventional

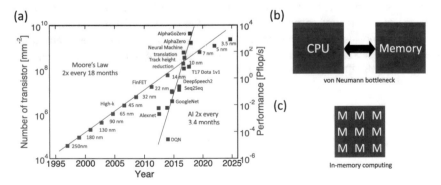

Fig. 6.1 **a** Comparison between the exponential growth of Moore's Law (red) and the required performance for executing modern artificial intelligence (AI) algorithms (blue), **b** conventional von-Neumann architecture suffering from a bottleneck when transferring data between memory and processing unit and **c** in-memory computing concept

von-Neumann architecture [22], where the processing unit (blue) is responsible only for executing operations whereas the memory unit (red) is responsible for storing them. Most of nowadays computers are based on this architecture, where one or multiple types of memory store the data with the central processing units (CPU) or graphic processing units (GPU) performing computation. When a lot of data need to be analyzed this architecture exposes a bottleneck in computation, known as von-Neumann bottleneck [19], due to the time and energy spent for handling data travelling from memory to processor and back. A new computing architecture that avoids the bottleneck is then desired.

In-memory computing with novel resistive memories [10, 11] has been proposed as a solution to overcome the limitation of both Moore's law speed and von Neumann bottleneck. The idea is to harness intrinsic materials properties of such memories to create new computing paradigms based on physical laws and known as physical computing [11, 44]. By organizing memories in crosspoint arrays it is possible then to have a compact accelerator known as memory processing unit (MPU) [44], which does not require data transfer and can perform computations within the memory. Figure 6.1c shows a conceptual representation of a MPU architecture, with many memory cores interconnected with each other. The novel computing unit have been shown to have unprecedented speed up compared with traditional and specific circuit for acceleration [25].

6.3 In-Memory Matrix-Vector-Multiplication Accelerator

Emerging memory devices, commonly referred as memristors, have recently attracted interest for their application both as memory and computing elements. Among these, resistive random access memories (RRAM) are a promising candidate for computing, due to their low energy operation, high endurance, small area and cost-effective fabrication [9]. Figure 6.2a shows a typical current-voltage characteristic of a RRAM device which is depicted in the inset and made of a Ti top electrode (TE) deposited on a HfO_x layer and a C bottom electrode (BE) [2]. After a forming process it is possible to change and modulate the conductance of the device. A positive pulse applied from the TE to the BE will result in a filament growth from TE to BE, or set transition, bringing the device into a low resistance state (LRS). By fixing the maximum current flowing to the RRAM during the set transition, namely compliance current (I_C), it is possible to avoid hard breaks of the device oxide and modulate the LRS conductance. I_C can be fixed by an external circuit, such a Source Measurement Unit (SMU), or with a transistor connected with the drain at the BE, that can also be used as selector device in an array configuration. By applying a negative pulse the RRAM undergoes a reset, resulting in the filament rupture and a gap formation in the conductive path, thus an high resistance state (HRS). The gap width can be controlled by the maximum applied negative voltage during reset and can be used as well to modulate the conductance. Figure 6.2b shows different measured conductance states demonstrating the possibility of analog programming of the RRAM device. Given

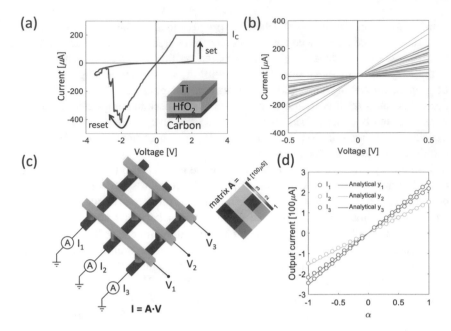

Fig. 6.2 **a** Resistive Random Access Memory (RRAM) I-V characteristics. By applying a positive voltage it is possible to set the memory device into a low resistance state (LRS) whose conductance is controlled by the maximum compliance current I_C flowing during the set operation, while by applying a negative voltage it is possible to reset the device into a high resistance state (HRS) whose depth is controlled by the maximum applied negative voltage. Inset shows the fabricated Ti-HfO$_x$-C RRAM device. **b** Different conductance achieved by modulating I_C during the set operation. **c** Crosspoint memory architecture for multiply-accumulate operation. RRAM devices are organized in an array representing an analog matrix A, by applying a voltage vector V on the columns, the current vector I at the rows is the matrix-vector multiplication result $I = GV$. Inset represent a measured programmed matrix A. **d** Measured (circles) and calculated (lines) currents vector I as function of the parameter α controlling the applied voltage vector $V = \alpha[0.2, 0.3, 0.4]$ with $-1 \le \alpha \le 1$. Adapted from [33]

the possibility of representing in principle any given number, applications in analog operation acceleration with RRAM devices have rapidly arisen. Different architectures have been presented to accelerate analog problems such as crosspoint arrays [40] and content addressable memories [17]. Figure 6.2c shows a crosspoint array implementation where memristive devices are arranged in a matrix form to directly write an algebraic matrix of real positive numbers G into the RRAM conductance. By applying a input voltage vector V on the crosspoint columns, the current flowing through the crosspoint rows is $I = GV$ or the dot product of matrix G by vector V. In this way, it is possible to accelerate dot product, also referred as matrix-vector-multiplication (MVM), in one step [10, 11]. Memristive crosspoint has been shown able to accelerate different problems based on MVM, such as the training [16, 27, 38] and inference [20, 41] of neural networks, image processing [18], sparse coding [29], optimization problems [6, 24] and the solution of linear equations through iterative numerical approaches [14, 43]. Integrated circuits comprising memristive arrays and the circuitry need to generate the input, such as digital-analog-converters (DAC), sense and read the outputs, such as transimpedance amplifiers (TIA) and analog-digital-converters (ADC), and cell selecting and routing, able to accelerate MVM have been proposed [5, 37, 42], outperforming modern processor both in throughput and energy saving [42].

6.4 One Step in-Memory Solution of Inverse Algebraic Problems

Crosspoint arrays offer the analog capability of writing arbitrary positive real matrixes coefficients, however iterative operations are usually performed in conventional digital hardware [6, 43]. To harness the full potential of the analog approach, iterations can be performed in the analog domain through feedback connected operational amplifiers [23, 33, 34]. By properly programming the conductance matrix and connecting the feedback amplifiers, one can solve different inverse problems such as linear systems [33], eigenvectors calculation and pageranking[32], linear and logistic regressions [34].

6.4.1 In-Memory Solution of Linear Systems in One-Step

Operational amplifiers in negative feedback configuration offer analog implementation of loops. Solving a system of linear equation is the equivalent matrix operation of performing a division between two scalars. This is the role of a TIA, an operational amplifier with a feedback resistance R connected between the negative input and the output. Grounding the positive input and injecting a current I on the negative input, the output voltage will adjust on $V = IR$ or $V = I/G$ with $G = 1/R$ conductance of the resistance R. This is due to the negative feedback effect and the nature of the operational amplifier that has a very large input impedance. By considering a matrix version of this circuit, it is then possible to calculate the solution of a linear system encoded in a matrix of conductance G, which is connected in feedback with operational amplifiers [33]. Figure 6.3a shows the circuit schematic for a 3 equations linear system. The system coefficients are encoded in the conductance matrix A (Fig. 6.3a inset) measured on 9 HfO$_x$ RRAM devices arranged in crosspoint configuration. The crosspoint rows are connected to the negative input of the operational amplifiers, the columns to the output of the operational amplifiers while the positive input of the operational amplifiers is kept connected to ground. By injecting a current I on the rows representing the known vector of the linear system the output voltage vector will be the solution of the linear system $V = A^{-1}I$, which is computed in one step without digital iteration [33]. Figure 6.3b demonstrate the concept showing

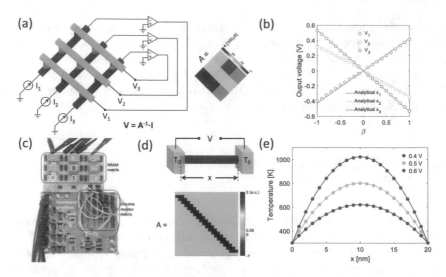

Fig. 6.3 **a** Circuit for solving a linear system in one step comprising a cross-point array of RRAM devices (red cylinders) programmed with the conductance matrix A (inset), connected with the rows (blue bars) at the negative input of operational amplifiers. By injecting a current I through the rows, the columns (green lines) connected to the operational amplifier outputs will stabilized to a voltage vector $V = A^{-1}I$ which is the solution of a linear system. **b** Measurements of output voltages (circles) and analytical (lines) solution of the linear systems $AV = I$ with $I = \beta[0.2; 1; 1]$ and $-1 \leq \beta \leq 1$ as function of the controlling parameter β. The comparison of the measured voltages with the analytical solution support the accuracy of the system. **c** Hardware implementation of the circuit for solving linear systems with HfO$_x$ devices arranged on a printed circuit board with commercial operational amplifiers controlled with an external arbitrary waveform generator. **d** 1-dimensional Fourier equation encoded with 21 points in a 21 × 21 conductance matrix. **e** Output voltages V (circle) simulated with SPICE for a larger circuit showing the solution of the 1-D Fourier heat equation confirming a good agreement with the analytical result (lines). Adapted from [33]

the measured voltage V and the analytical solution of the linear system $AV = I$ with $I = \beta[0.2; 1; 1]$ as function of the controlling parameter β, indicating a good agreement between electrical measurements and analytical result. The measurements were performed on a printed circuit board (PCB) with Ti/HfO$_x$/C RRAM devices [2] arranged on a crosspoint configuration and connected in feedback with commercial operational amplifiers (Fig. 6.3c). The known vector is given as voltage with an arbitrary waveform generator and then converted to current with input resistance connected to the negative input of the operational amplifiers. The output voltage is monitored with an external oscilloscope. To represent both positive and negative coefficients of the linear system, it is possible to use two separated crosspoint that represent the matrixes B and C, with $A = B - C$. By connecting matrix B to the circuit of Fig. 6.3a, the output voltage to the matrix C through negative buffers and feeding both matrix B and matrix C with the same input current representing the known vector, one can solve an arbitrary linear system $(B - C)V = AV = I$ where A has both positive, negative and zero coefficients [33]. As an example, this circuit can be used to solve differential equations such as the Fourier heat equation [33]. Figure 6.3d shows a 1-D Fourier heat equation encoded in its 21×21 discretized matrix form, that can directly be mapped in a crosspoint array and be solved in one step. Figure 6.3e shows the output voltage vector V simulated with in SPICE representing the solution of the problem in Fig. 6.3d for different starting temperature compared with the analytical results and as function of the distance.

The results shows a good match supporting the use of the circuit for solving large systems of equations. In fact, interestingly the solution time does not depends on the matrix, thus linear system, dimension [35]. One can think about the operational amplifier in negative feedback configuration, where the bandwidth is limited by the loop gain and equal to $f_{max} = GBWP \cdot \frac{R_{in}}{R_{in}+R_f}$ where $GBWP$ is the gain bandwidth product of the operational amplifiers, R_{in} the input resistance and R_f the feedback resistance. By considering a feedback matrix, it is possible to demonstrate that the settling time, thus the bandwidth, is solely limited by the minimal eigenvalue of the matrix [35] and not by its size making the time complexity $O(1)$. This is an unprecedented speedup compared with conventional conjugate gradient solvers [30], where time complexity is $O(N)$ at its best and quantum computing [8], where the best time complexity is $O(log(N))$ where N is the size (i.e. the number of equations) in the linear system. The result supports the use of the circuit for solving systems of linear equations in one step, outperforming digital and quantum computers.

6.4.2 In-Memory Eigenvectors Calculation in One-Step

Many scientific and machine learning problems, such as the solution of differential equations, require not the simple solution of a linear system, but the eigenvector computation. Mathematically speaking, this means to solve the equation $Ax = \lambda x$, which can be arranged such as $(A - \lambda I)x = 0$. It is possible to observe that by encoding on a crosspoint the matrix A and on a second crosspoint the diagonal matrix λI, with the mixed matrix configuration it is possible to compute the eigenvectors with the feedback circuit of Sect. 6.4.1 [32, 33]. Figure 6.4a shows a compact circuit schematic for calculating the eigenvector solution where the diagonal matrix is represented with feedback conductance G_λ. To guarantee the stability of the circuit, only the eigenvectors corresponding to highest positive and lower negative eigenvalue can be computed. In fact the circuit works at the boundary of stability with a loop gain $G_{Loop} = 1$. Without any input current the opamp corresponding to the maximum value of the eigenvector saturates while the others adjust resulting in an output voltage vector V, which normalized by the supply voltage, it is equal to the normalized eigenvector x corresponding to the non-trivial solution of $(A - \lambda I)x = 0$. Figure 6.4a-inset shows a programmed conductance matrix A and Fig. 6.4b the measured eigenvectors calculation corresponding to the highest positive (red) and lowest negative (blue) eigenvalue as function of the analytical solution, showing a good agreement. It has to be noted that to compute the eigenvector corresponding to the negative eigenvalue

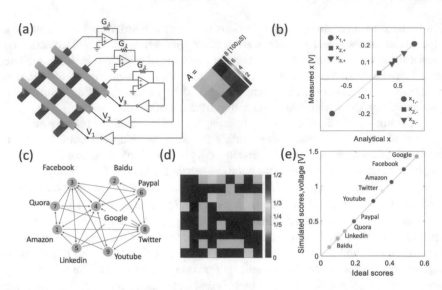

Fig. 6.4 **a** Circuit for solving the eigenvector equation $Ax = \lambda x$, where x is the eigenvector and λ the maximum positive eigenvalue. With an input current $I = 0$, the operational amplifier corresponding to the maximum value of the eigenvector x saturates. By normalizing the output voltages, the eigenvector is found. Inset shows a 3×3 matrix encoded in RRAM conductance. **b** Experimental solution of the eigenvector corresponding to the highest positive (red) and lowest negative (blue) eigenvalue, as function of the analytical solution. The eigenvalues are encoded in the feedback conductance G_λ. **c** Illustration of Pagerank algorithm, web pages are represented by green circles and the corresponding citation with blue arrows. **d** Stochastic link matrix corresponding to the problem in (**c**), which is calculated by normalizing the boolean link matrix by the sum over each column. **e** Simulation (circles) results of Pagerank problem in (**c**) as function of the ideal ranking. Adapted from [33]

the analog inverter of Fig. 6.4 should be removed with the output voltages of the operational amplifiers directly connected to the crosspoint array A. Unfortunately, in any case the highest eigenvalue λ_1 must be known. To do that it is possible to apply iterative solution such as power iteration, or a sweep the conductance G_λ until one of the operational amplifier saturates. However, for some applications such as Pagerank the algorithm at the backbone of Google search engine [4], the maximum eigenvalue is always known a priori. Figure 6.4c shows an illustration of a web pages network with pages represented with green circles and citation represented by blue arrows. Goal of pageranking is to give a score to every webpage corresponding to its authority, namely how many citation receives from other pages with high authority. To do that it is possible to compute the eigenvector corresponding to the maximum eigenvalue of a stochastic matrix, namely the boolean link matrix between webpages normalized by the sum over each column [32, 33]. Interestingly, the maximum eigenvalue of such matrix is always known and $\lambda_1 = 1$, making the system highly feasible for giving such solution. Figure 6.4d shows the stochastic matrix corresponding to the network in Fig. 6.4c, whose SPICE simulated eigenvector solution is plotted in

Fig. 6.4e as function of the ideal solution showing good agreement. The circuit was also simulated with measured RRAM conductance tuned with a program and verify algorithm showing good agreement with the Hardvard500 dataset results [32]. As the circuit in Sect. 6.4.1, the circuit for eigenvector computation shows a constant time complexity $O(1)$ [32], making it aggressively interesting for machine learning and scientific applications compared with other computing technologies.

6.4.3 In-Memory Regression and Classification in One-Step

Many computing problems have more unknowns than equations or more equations than unknowns. The latter is the case of regression problem, which is a fundamental machine learning model for predicting a certain data behavior or classify its class. Linear and logistic regression are among the most used ML algorithms [1]. A linear regression problem can be described with the overdetermined linear system $Xw = y$, where X is a $N \times M$ matrix with $N > M$, y is known vector of size $N \times 1$ and w is the unknown weight vector of size $M \times 1$. There is no exact solution to this problem, but the best solution can be calculated with the least squares error approach, that minimizes $||\epsilon|| = ||Xw - y||_2$ which is the euclidean norm of the error. This can be done through the Moore-Penrose pseudoinverse [26] solving the equation

$$w = (X^T X)^{-1} X^T y. \tag{6.1}$$

To calculate w is one step, it is possible to cascade multiple analog stages representing all the parts of the equation. Figure 6.5a shows a schematic of the realized circuit for calculating linear regression weights in one step [34]. The conductance matrix X encodes the explanatory variables while the dependent variables are injected as current I. The output voltage of the rows amplifier will then adjust on $V_{row} = (VX + I)/G_{TI}$, thanks to the transimpedance configuration. Being the columns of the right crosspoint array connected to the input of the columns operational amplifier the current should be equal to zero, such as

$$\frac{(VX + I)}{G_{TI}} X^T = 0. \tag{6.2}$$

By rearranging equation (6.2), it is possible to observe that the weights of equation (6.1) are obtained in one step, without iterations as voltage V [34]. The inset of Fig. 6.5a shows a programmed conductance matrix on HfO_x arranged in a double array configuration and representing the linear regression problem of Fig. 6.5b, which shows a comparison of the experimental linear regression and the analytical one, evidencing a good agreement. Interestingly, with the same circuit is also possible to compute logistic regression in one step, thus classify data. By encoding in the conductance matrix the explanatory variables and injecting the class as input current, indeed it is possible to obtain the weights corresponding to a binary classification of

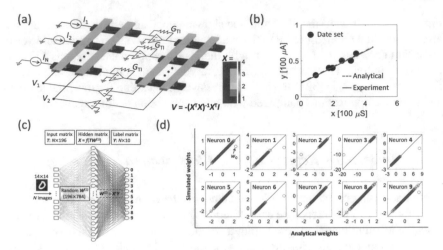

Fig. 6.5 **a** Circuit for calculating regressions operation trough the Moore-Penrose pseudo inverse. Inset shows a programmed linear regression problem. **b** Experimental results and analytical calculation of the linear regression of 6 data points. **c** Neural network topology implemented for the weights optimization in one step. **d** Simulated weights as function of the analytical weights for the training of the neural network classification layer. Adapted from [34]

data. To illustrate such concept it is possible for example to train an output layer of a neural network in one step. Figure 6.5c shows a neural network topology, namely an extreme learning machine (ELM) used as example for neural network training. The network is made of 196 input neurons (corresponding to the pixels of an input image from the MNIST dataset reshaped on a 14×14 size), 784 hidden neurons on a single hidden layer and 10 output neurons corresponding to the numbers from 0 to 9 of the MNIST dataset [15]. The first layer weights are randomized with a uniform distribution between 1 and -1 and the output last layer is trained with logistic regression. By encoding in the conductance matrix the dataset evaluated on the hidden layer it is possible to use the circuit for calculating the weights of the second layer corresponding to a single output neuron in one step [34]. Figure 6.5d shows a comparison between the analytical weights and the simulated weights with a SPICE circuit simulation, showing little differences. The accuracy of the network trained with the circuit in recognizing the MNIST dataset is 92% which is equivalent to the ideal result for such network.

To evaluate the performance of the circuit it is possible to consider the number of computing steps required for training such neural network on a von Neumann architecture. With conventional computing approach, the complexity for calculating the logistic regression weights of equation (6.1) is composed by $O(M^2 N)$ to multiply X^T by X, $O(MN)$ to multiply X^T by y and $O(M^3)$ to compute the LU factorization of XX^T and use it to calculate $(XX^T)^{-1}$. Thus, $M^2 N + MN + M^3$ floating points operations are required. In the case of the training of the neural network classification layer of Fig. 6.5c, 2.335×10^9 operations are required. Given that the simulated weight training with the in-memory closed loop crosspoint circuit required 145 us [34], the circuit has an equivalent throughput of 16.1 TOPS. The overall power consumption of the simulated circuit is calculated to be 355.6 mW [34] per training operation assuming a conductance unit of $10\mu S$ in the circuit. As a result the efficiency of the circuit is calculated to be 45.3 TOPS/W. As an approximate comparison the energy efficiency of Google TPU is 2.3 TOPS/W [12] while the energy efficiency of an in-memory open loop circuit is 7.02 TOPS/W [29], evidencing that the in-memory closed loop solution is 19.7 and 6.5 times more efficient, respectively. The results show the appealing feasibility of the in-memory computing circuit for solving machine learning tasks, such as training a neural network with unprecedented throughput.

6.5 Conclusions

In this chapter in-memory circuit accelerators for inverse algebra problems have been presented. Compared to previous results, thanks to operational amplifiers connected in feedback configuration, it is possible to solve such problems in just one step. First the open loop crosspoint circuit for matrix vector multiplication is illustrated. Then, the novel crosspoint closed loop circuits are demonstrated able of solving linear systems and computing eigenvectors, in one step without iterations. Finally, the concept is extended to machine learning tasks such as linear regression and neural networks training in one step. These results supports in-memory computing as a future computing paradigm to obtain size independent time complexity solution of algebraic problems in a compact and low energy platform.

Acknowledgements The author would like to thank P. Mannocci for his critical reading of the manuscript.

References

1. The State of Data Science and Machine Learning (2017). https://www.kaggle.com/surveys/2017
2. Ambrosi E, Bricalli A, Laudato M, Ielmini D (2019) Impact of oxide and electrode materials on the switching characteristics of oxide ReRAM devices. Faraday Discuss 213:87–98. https://doi.org/10.1039/C8FD00106E
3. Amodei D, Hernandez D. AI and compute. https://openai.com/blog/ai-and-compute/
4. Bryan K, Leise T (2006) The $25,000,000,000 eigenvector: the linear algebra behind google. SIAM Rev 48(3):569–581. https://doi.org/10.1137/050623280
5. Cai F, Correll JM, Lee SH, Lim Y, Bothra V, Zhang Z, Flynn MP, Lu WD (2019) A fully integrated reprogrammable memristor-CMOS system for efficient multiply-accumulate operations. Nat Electron 2(7):290–299. https://doi.org/10.1038/s41928-019-0270-x
6. Cai F, Kumar S, Vaerenbergh TV, Liu R, Li C, Yu S, Xia Q, Yang JJ, Beausoleil R, Lu W, Strachan JP (2019) Harnessing intrinsic noise in memristor hopfield neural networks for combinatorial optimization. https://arxiv.org/1903.11194

7. Golub GH, Van Loan CF (2013) Matrix computations, 4th edn. Johns Hopkins studies in the mathematical sciences. The Johns Hopkins University Press, Baltimore. OCLC: ocn824733531
8. Harrow AW, Hassidim A, Lloyd S (2009) Quantum algorithm for linear systems of equations. Phys Rev Lett 103(15):150502. https://doi.org/10.1103/PhysRevLett.103.150502
9. Ielmini D (2016) Resistive switching memories based on metal oxides: mechanisms, reliability and scaling. Semicond Sci Technol 31(6):063002. https://doi.org/10.1088/0268-1242/31/6/063002
10. Ielmini D, Pedretti G (2020) Device and circuit architectures for in-memory computing. Adv Intell Syst, p 2000040. https://doi.org/10.1002/aisy.202000040
11. Ielmini D, Wong HSP (2018) In-memory computing with resistive switching devices. Nat Electron 1(6):333–343. https://doi.org/10.1038/s41928-018-0092-2
12. Jouppi NP, Borchers A, Boyle R, Cantin Pl, Chao C, Clark C, Coriell J, Daley M, Dau M, Dean J, Gelb B, Young C, Ghaemmaghami TV, Gottipati R, Gulland W, Hagmann R, Ho CR, Hogberg D, Hu J, Hundt R, Hurt D, Ibarz J, Patil N, Jaffey A, Jaworski A, Kaplan A, Khaitan H, Killebrew D, Koch A, Kumar N, Lacy S, Laudon J, Law J, Patterson D, Le D, Leary C, Liu Z, Lucke K, Lundin A, MacKean G, Maggiore A, Mahony M, Miller K, Nagarajan R, Agrawal G, Narayanaswami R, Ni R, Nix K, Norrie T, Omernick M, Penukonda N, Phelps A, Ross J, Ross M, Salek A, Bajwa R, Samadiani E, Severn C, Sizikov G, Snelham M, Souter J, Steinberg D, Swing A, Tan M, Thorson G, Tian B, Bates S, Toma H, Tuttle E, Vasudevan V, Walter R, Wang W, Wilcox E, Yoon DH, Bhatia S, Boden N (2017) In-datacenter performance analysis of a tensor processing unit. In: Proceedings of the 44th annual international symposium on computer architecture - ISCA '17, pp 1–12. ACM Press, Toronto, ON, Canada. https://doi.org/10.1145/3079856.3080246
13. Landauer R (1988) Dissipation and noise immunity in computation and communication. Naure 335(27):779–784
14. Le Gallo M, Sebastian A, Mathis R, Manica M, Giefers H, Tuma T, Bekas C, Curioni A, Eleftheriou E (2018) Mixed-precision in-memory computing. Nat Electron 1(4):246–253. https://doi.org/10.1038/s41928-018-0054-8
15. Lecun Y, Bottou L, Bengio Y, Haffner P (1998) Gradient-based learning applied to document recognition. In: Proceedings of the IEEE 86(11):2278–2324. https://doi.org/10.1109/5.726791
16. Li C, Belkin D, Li Y, Yan P, Hu M, Ge N, Jiang H, Montgomery E, Lin P, Wang Z, Song W, Strachan JP, Barnell M, Wu Q, Williams RS, Yang JJ, Xia Q (2018) Efficient and self-adaptive in-situ learning in multilayer memristor neural networks. Nat Commun 9(1):2385. https://doi.org/10.1038/s41467-018-04484-2
17. Li C, Graves CE, Sheng X, Miller D, Foltin M, Pedretti G, Strachan JP (2020) Analog content-addressable memories with memristors. Nat Commun 11(1):1638. https://doi.org/10.1038/s41467-020-15254-4
18. Li C, Hu M, Li Y, Jiang H, Ge N, Montgomcry E, Zhang J, Song W, Davila N, Graves CE, Li Z, Strachan JP, Lin P, Wang Z, Barnell M, Wu Q, Williams RS, Yang JJ, Xia Q (2018) Analogue signal and image processing with large memristor crossbars. Nat Electron 1(1):52–59. https://doi.org/10.1038/s41928-017-0002-z
19. Merolla PA, Arthur JV, Alvarez-Icaza R, Cassidy AS, Sawada J, Akopyan F, Jackson BL, Imam N, Guo C, Nakamura Y, Brezzo B, Vo I, Esser SK, Appuswamy R, Taba B, Amir A, Flickner MD, Risk WP, Manohar R, Modha DS (2014) A million spiking-neuron integrated circuit with a scalable communication network and interface. Science (6197):668–673. https://doi.org/10.1126/science.1254642
20. Milo V, Zambelli C, Olivo P, Perez E, Mahadevaiah MK, Ossorio OG, Wenger C, Ielmini D (2019) Multilevel HfO$_2$ -based RRAM devices for low-power neuromorphic networks. APL Mater 7(8):081120. https://doi.org/10.1063/1.5108650
21. Moore GE (2006) Cramming more components onto integrated circuits, Reprinted from Electronics, volume 38, number 8, April 19, 1965, pp.114 ff. IEEE Solid-State Circuits Soc Newsl 11(3): 33–35. https://doi.org/10.1109/N-SSC.2006.4785860
22. von Neumann J (1945) First draft of a report on the EDVAC. https://doi.org/10.5555/1102046

23. Pedretti G (2020) In-memory computing with memristive devices. Ph.D. thesis, Politecnico di Milano
24. Pedretti G, Mannocci P, Hashemkhani S, Milo V, Melnic O, Chicca E, Ielmini D (2020) A spiking recurrent neural network with phase change memory neurons and synapses for the accelerated solution of constraint satisfaction problems. IEEE J Explor Solid-State Comput Devices Circuits, pp 1–1. https://doi.org/10.1109/JXCDC.2020.2992691. https://ieeexplore. ieee.org/document/9086758/
25. Peng X, Kim M, Sun X, Yin S, Rakshit T, Hatcher RM, Kittl JA, Seo JS, Yu S (2019) Inference engine benchmarking across technological platforms from CMOS to RRAM. In: Proceedings of the international symposium on memory systems - MEMSYS '19, pp 471–479. ACM Press, Washington, District of Columbia. https://doi.org/10.1145/3357526.3357566
26. Penrose R (1955) A generalized inverse for matrices. Math Proc Camb Philos Soc 51(3):406–413. https://doi.org/10.1017/S0305004100030401
27. Prezioso M, Merrikh-Bayat F, Hoskins BD, Adam GC, Likharev KK, Strukov DB (2015) Training and operation of an integrated neuromorphic network based on metal-oxide memristors. Nature 521(7550):61–64. https://doi.org/10.1038/nature14441
28. Salahuddin S, Ni K, Datta S (2018) The era of hyper-scaling in electronics. Nat Electron 1(8):442–450. https://doi.org/10.1038/s41928-018-0117-x
29. Sheridan PM, Cai F, Du C, Ma W, Zhang Z, Lu WD (2017) Sparse coding with memristor networks. Nat Nanotechnol 12(8):784–789. https://doi.org/10.1038/nnano.2017.83
30. Shewchuk JR (1994) An introduction to the conjugate gradient method without the agonizing pain. Technical report CMU-CS-94-125, School of Computer Science, Carnegie Mellon University, Pittsburgh
31. Sun Z, Ambrosi E, Bricalli A, Ielmini D (2018) logic computing with stateful neural networks of resistive switches. Adv Mater 30(38):1802554. https://doi.org/10.1002/adma.201802554
32. Sun Z, Ambrosi E, Pedretti G, Bricalli A, Ielmini D (2020) In-memory pagerank accelerator with a cross-point array of resistive memories. IEEE Trans Electron Devices 67(4):1466–1470. https://doi.org/10.1109/TED.2020.2966908. https://ieeexplore.ieee.org/document/8982173/
33. Sun Z, Pedretti G, Ambrosi E, Bricalli A, Wang W, Ielmini D (2019) Solving matrix equations in one step with cross-point resistive arrays. Proc Natl Acad Sci 116(10):4123–4128. https:// doi.org/10.1073/pnas.1815682116
34. Sun Z, Pedretti G, Bricalli A, Ielmini D (2020) One-step regression and classification with cross-point resistive memory arrays. Sci Adv 6(5):caay2378. https://doi.org/10.1126/sciadv. aay2378
35. Sun Z, Pedretti G, Mannocci P, Ambrosi E, Bricalli A, Ielmini D (2020) Time complexity of in-memory solution of linear systems. IEEE Trans Electron Devices, pp 1–7. https://doi.org/ 10.1109/TED.2020.2992435. https://ieeexplore.ieee.org/document/9095220/
36. Tan L, Kothapalli S, Chen L, Hussaini O, Bissiri R, Chen Z (2014) A survey of power and energy efficient techniques for high performance numerical linear algebra operations. Parallel Comput 40(10):559–573. https://doi.org/10.1016/j.parco.2014.09.001. https://linkinghub. elsevier.com/retrieve/pii/S0167819114001112
37. Wan W, Kubendran R, Eryilmaz SB, Zhang W, Liao Y, Wu D, Deiss S, Gao B, Raina P, Joshi S, Wu H, Cauwenberghs G, Wong HSP (2020) 33.1 A 74 TMACS/W CMOS-RRAM neurosynaptic core with dynamically reconfigurable dataflow and in-situ transposable weights for probabilistic graphical models. In: 2020 IEEE international solid- state circuits conference - (ISSCC), pp 498–500. IEEE, San Francisco, CA, USA. https://doi.org/10.1109/ISSCC19947. 2020.9062979
38. Wang Z, Li C, Lin P, Rao M, Nie Y, Song W, Qiu Q, Li Y, Yan P, Strachan JP, Ge N, McDonald N, Wu Q, Hu M, Wu H, Williams RS, Xia Q, Yang JJ (2019) In situ training of feed-forward and recurrent convolutional memristor networks. Nat Mach Intell 1(9):434–442. https://doi. org/10.1038/s42256-019-0089-1
39. Wang Z, Wu H, Burr GW, Hwang CS, Wang KL, Xia Q, Yang JJ (2020) Resistive switching materials for information processing. Nat Rev Mater. https://doi.org/10.1038/s41578-019-0159-3

40. Yang JJ, Strukov DB, Stewart DR (2013) Memristive devices for computing. Nat Nanotechnol 8(1):13–24. https://doi.org/10.1038/nnano.2012.240
41. Yao P, Wu H, Gao B, Eryilmaz SB, Huang X, Zhang W, Zhang Q, Deng N, Shi L, Wong HSP, Qian H (2017) Face classification using electronic synapses. Nat Commun 8(1):15199. https://doi.org/10.1038/ncomms15199
42. Yao P, Wu H, Gao B, Tang J, Zhang Q, Zhang W, Yang JJ, Qian H (2020) Fully hardware-implemented memristor convolutional neural network. Nature 577(7792):641–646 (2020). https://doi.org/10.1038/s41586-020-1942-4
43. Zidan MA, Jeong Y, Lee J, Chen B, Huang S, Kushner MJ, Lu WD (2018) A general memristor-based partial differential equation solver. Nat Electron 1(7):411–420. https://doi.org/10.1038/s41928-018-0100-6
44. Zidan MA, Strachan JP, Lu WD (2018) The future of electronics based on memristive systems. Nat Electron 1(1):22–29. https://doi.org/10.1038/s41928-017-0006-8

Open Access This chapter is licensed under the terms of the Creative Commons Attribution 4.0 International License (http://creativecommons.org/licenses/by/4.0/), which permits use, sharing, adaptation, distribution and reproduction in any medium or format, as long as you give appropriate credit to the original author(s) and the source, provide a link to the Creative Commons license and indicate if changes were made.

The images or other third party material in this chapter are included in the chapter's Creative Commons license, unless indicated otherwise in a credit line to the material. If material is not included in the chapter's Creative Commons license and your intended use is not permitted by statutory regulation or exceeds the permitted use, you will need to obtain permission directly from the copyright holder.

Chapter 7
Development of a 3" LaBr3 SiPM-Based Detection Module for High Resolution Gamma Ray Spectroscopy and Imaging

Giovanni Ludovico Montagnani

7.1 Introduction

My thesis work entails the design of each element of the instrument (Fig. 7.1). Starting from the SiPM tile and the optimization of the instrument mechanics, through to the development of the electronics boards and custom ASICs. The Gain Amplitude Modulated Multichannel ASIC (GAMMA) was developed in order to match the project requirements for a large charge dynamic range and full scale range. The ASIC is meant to cope with the 14 dB charge range provided by the SIPMs and to provide this information minimizing the statistical contribution to resolution degradation. The exploitation of a Gated Integrator filter with self triggering capabilities was meant to optimize the signal collection versus the integration of SiPM dark current. During the years of work, me and my team of students, developed a first 8-channel ASIC, and a second release exploiting 16 analog channels and multiplexers. The ASIC main schematic is represented by Fig. 7.2: the core of the chip is the 16 channels stack which exploits a current input stage, the Gated Integrator with Track and Hold feature, an Active Gain Control and a Baseline Holder circuit [2].

7.2 Development

A peculiar feature of the GAMMA ASIC is the Active Gain Change mechanism, that profit from a "predictive" analog algorithm to choose the best gain of the amplifier to process the input pulse in order to minimise the contribution of the output noise as depicted by Fig. 7.3. Another important feature of the analog filter is the Baseline

G. L. Montagnani (✉)
Dipartimento di Elettronica, Informazione e Bioingegneria, Politecnico di Milano, Piazza
Leonardo da Vinci 32, Milano, Italy
e-mail: giovanni.montagnani@polimi.it

© The Author(s) 2021
A. Geraci (ed.), *Special Topics in Information Technology*,
PoliMI SpringerBriefs, https://doi.org/10.1007/978-3-030-62476-7_7

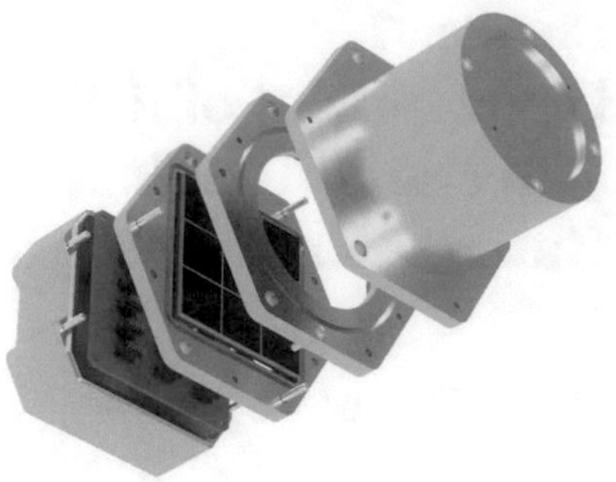

Fig. 7.1 Developed spectrometer enclosure. The cylindric shape is designed to contain the crystal, the hexagonal flange allows the exploitation of the system in laboratory setups and the upper box hosts the electronics

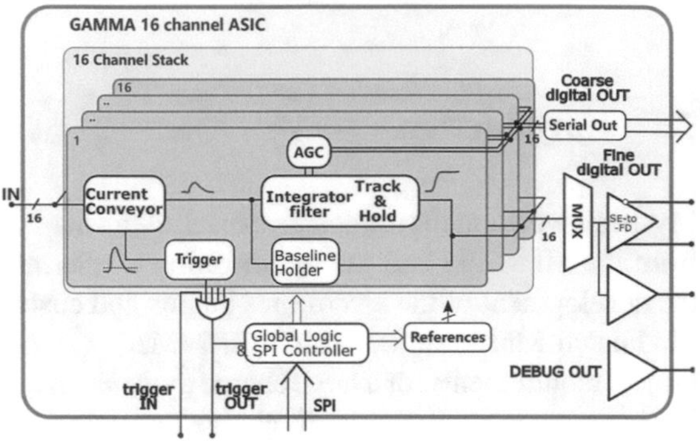

Fig. 7.2 GAMMA ASIC main schematic. The analog channel exploits a current input stage, the Gated Integrator with Track and Hold feature, an Active Gain Control and a Baseline Holder circuit

Fig. 7.3 Active Gain Change time diagram. The ASIC swaps the integrating capacitance during the integration time sensing the output voltage

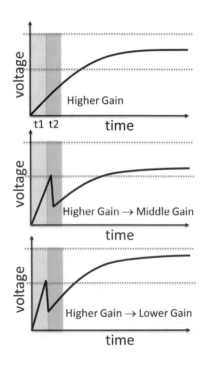

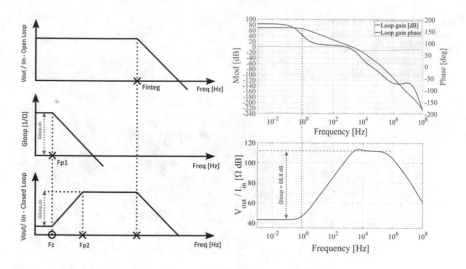

Fig. 7.4 Baseline Holder filter working principle. By placing a very low frequency low pass filter in feedback to the Gated Integrator a bandpass effect is obtained

Holder used to avoid DC current integration. This 1 Hz low pass filter with non-linear behaviour is used in feedback to the Gated Integrator filter, in order to drain the DC current from the input stage as depicted by Fig. 7.4. The non-linearity is useful to avoid a dependence of the subtracted DC current from the amplitude of the current pulses, but it must be dimensioned carefully in order to avoid unwanted instability issues [1]. The final 16 channel ASIC also hosts an SPI programmable memory to tune the filters parameters trough 7 internal DACs and a differential output buffer to match the requirements of the external 14 differential ADC. The ASIC and the ADC are placed on a small PCB module to be placed on a larger motherboard for improved connectivity and modularity [3].

Data acquisition and the biasing of the detectors were also designed in order to simplify the use of the instrument in operative conditions. An important feature that allowed state of the art results is the SiPM gain stabilization through overvoltage correction versus temperature variation. An ARTIX-7 FPGA, handles the acquisition from the ASICs, data storage, processing, reconstruction algorithm and USB 2.0 communication [5] (Fig. 7.5).

Experimental measurements were performed in intermediate development steps, confirming the high performance of the developed instrument. Considering that the main motivation of this work was to develop the first instrument capable of detecting rays from hundreds of thousands of electronvolts to tens of millions of electronvolts with high efficiency and state of the art energy resolution, the large number of active channels (120) was mandatory to achieve both spatial sensitivity and energy resolution. The measurements performed on a preliminary detection module have demonstrated the ability of the system to consistently achieve results in line with expectations—even better in some cases—within the restrictions given by the lim-

Fig. 7.5 SiPM readout and biasing circuitry. The system hosts 9 GAMMA ASICs, DC/DC regulators and FPGA for post-processing

ited number of readout channels and by the simplified microcontroller-based DAQ. A record energy resolution of 2.58% FWHM at 662 keV [4] has been achieved coupling the designed detector with a co-doped LaBr3 crystal by Saint-Gobain as reported in Figs. 7.6 and 7.7. This is, to the author's knowledge, the best resolution obtained coupling SiPM with scintillator crystals. A spatial resolution better than 10 mm at the center of the matrix has been obtained dividing the matrix in just 8 macro-pixels. The measurement dataset, shown in Fig. 7.8, was used to train an algorithm capable of reconstructing the horizontal position of interaction with sufficient precision. The next step of the project is to complete the 144-SiPM matrix readout system coupling each SiPM with a dedicated ASIC input: this is achieved by exploiting a total of 9 16-channels ASIC, of which a first release has been tested in parallel to this thesis and the results are reported in the dedicated chapter. The writing of this thesis occurred during the beginning of the testing phase. However, up to now, the only spectroscopic results available were obtained with single 8-channel or 16-channel ASICs in a reduced energy range. It will be possible to read individually the 120 SiPM of the matrix, increasing the energy full scale range to meet the project specification of 20 MeV. The combination of the preliminary system developed during this thesis work with the new 16 channels ASICs and the FPGA-based acquisition system will allow to obtain the ultimate performances targeted by the GAMMA project, satisfying all the requirements of the application: 120 non-merged SiPMs will allow to reach the target energy resolution while reaching the full required dynamic range, in conjunction with excellent spatial resolutions thanks to a large number of small pixels and to the promising neural-network-driven algorithm whose operation will be further refined. The new data acquisition system will permit to acquire data from a all the ASICs even at high event rates, together with an easy management of the SiPM bias voltage to compensate for gain variation due to temperature.

Future developments of my work could involve a better exploitation of the FPGA capability other than mere improvements of the developed hardware and

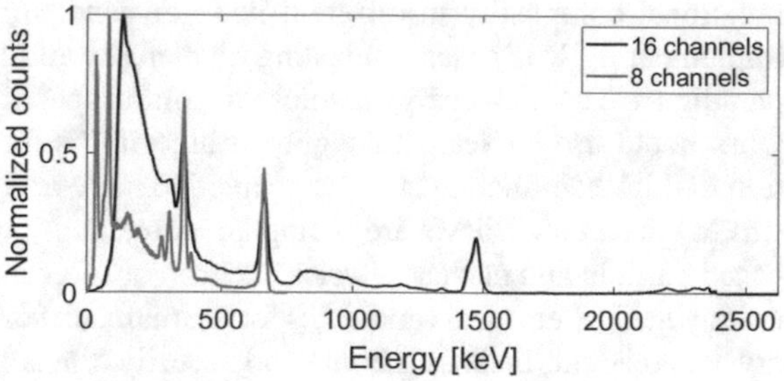

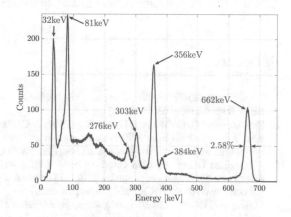

Fig. 7.6 Spectra comparison between the 8 channel and the 16 channel ASIC readout of the same 3" diameter LaBr3 crystal

Fig. 7.7 Acquired spectrum with record resolution of 2.58% @66 keV obtained from the readout of a 3" diameter LaBr3 Sr codoped crystal

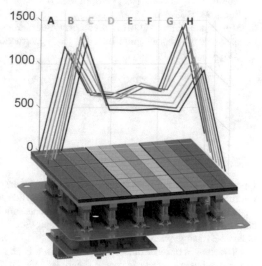

Fig. 7.8 Training dataset used for position reconstruction algorithm. Average signal distribution (vertical) for the 8 different positions corresponding to the centers (horizontal) of the pixels and interaction point reconstruction results using a cross-correlation-based algorithm

bug corrections. Availability of computationally efficient Machine Learning (ML) algorithms, combined to high performance, easily configurable FPGA and microcontrollers can push the state-of-the-art performances of modern embedded systems towards better results using new, distributed architectures. The sharing of computational complexity in different node of a network or electronic system when implementing ML poses interesting challenges and the need for new architecture paradigms. The high energy resolution gamma spectroscopy presented in this thesis work could use K-Nearest Neighbour algorithms for interaction point reconstruction in thick scintillation crystals coupled to large arrays of Silicon Photomultipliers (SiPM) detectors. SiPMs are seeing an increasingly widespread use in accelerator-based particle and nuclear physics experiment. Recent developments in SiPM technology made them ideal candidates for the many different applications of scintillation crystals readout, including homeland security, thanks to their large dynamic range, quantum efficiency, robustness and immunity to external magnetic fields.

References

1. Chen Y et al (2016) Stability of the baseline holder in readout circuits for radiation detectors. IEEE Trans Nucl Sci 63(1):316–324
2. De Geronimo G, O'Connor P, Grosholz J (1999) A CMOS baseline holder (BLH) for readout ASICs. In: 1999 IEEE nuclear science symposium. Conference record. 1999 nuclear science symposium and medical imaging conference (Cat. No. 99CH37019), vol 1. IEEE, pp 370–374
3. Montagnani GL et al (2018) GAMMASiPM ASIC: performance assessment and improved design with 87dB dynamic range. In: 2018 IEEE nuclear science symposium and medical imaging conference (NSS/MIC). IEEE [To be published]
4. Montagnani GL et al (2019) Spectroscopic performance of a Sr co-doped 3" LaBr3 scintillator read by a SiPM array. Nucl Instrum Methods Phys Res Sect A: Accel Spectrom Detect Assoc Equip 931:158–161
5. Montagnani GL et al (2019) A compact 4-decade dynamic range readout module for gamma spectroscopy and imaging. In: 2019 IEEE international symposium on circuits and systems (ISCAS). IEEE, pp 1–5

Open Access This chapter is licensed under the terms of the Creative Commons Attribution 4.0 International License (http://creativecommons.org/licenses/by/4.0/), which permits use, sharing, adaptation, distribution and reproduction in any medium or format, as long as you give appropriate credit to the original author(s) and the source, provide a link to the Creative Commons license and indicate if changes were made.

The images or other third party material in this chapter are included in the chapter's Creative Commons license, unless indicated otherwise in a credit line to the material. If material is not included in the chapter's Creative Commons license and your intended use is not permitted by statutory regulation or exceeds the permitted use, you will need to obtain permission directly from the copyright holder.

Part III
Computer Science and Engineering

Chapter 8
Velocity on the Web

Riccardo Tommasini

8.1 Introduction

The World Wide Web (Web) is a distributed system designed around a global naming system called URI [25]. Resources, the central abstraction, are not limited in their scope[1] and, although it is not mandatory, they are typically published along with a representation of their state. Using Hyper-Text Transfer Protocol (HTTP), Web applications, or more generically agents can access, exchange, and interact with resources' representations.

The decentralized nature of the Web makes it scalable but causes the spread of Data Variety [36]. Indeed, resources are heterogeneous and noisy. The Web of Data (WoD) is an extension of the Web that enables interoperability among Web applications encouraging data sharing. Semantic technologies like RDF, SPARQL, and OWL are the pillars of the WoD's technological stack.

From Smart Cities [27] to environmental monitoring [1], from Social Media analysis [31] to fake-news detection [32], a growing number of Web applications need to access and process data as soon as they arrive and before they are no longer valuable [16]. To this extent, the Web infrastructure is evolving,and new protocols are emerging, e.g., WebSockets and Server-Sent Events, and Application Public Interfaces (API) are becoming reactive, e.g., WebHooks.

In the big data context, the challenge mentioned above is known as Data Velocity [36], and Stream Processing (SP) is the research area that investigate how to handle it. SP solutions are designed to analyze *streams* and detect *events* [14] in real-time. However, SP technologies are inadequate to work on the Web. Data Velocity appears together with Data Variety and they lack flexible data models and expressive

---

[1]A *resource* is anything to which we can assign a Uniform Resource Identifier (URI).

R. Tommasini (✉)
University of Tartu, Narva Mnt 18, Tartu, Estonia
e-mail: riccardo.tommasini@ut.ee

© The Author(s) 2021
A. Geraci (ed.), *Special Topics in Information Technology*,
PoliMI SpringerBriefs, https://doi.org/10.1007/978-3-030-62476-7_8

manipulation languages that are necessary to handle heterogeneous data. On the other hand, semantic technologies are not designed for continuous and reactive processing of streams and events. Therefore, *Web applications cannot tame Data Velocity and Variety at the same time* [16].

Data Velocity and Variety affect the entire data infrastructure. On the Web, it means technologies for the *identification*, *representation*, the *interaction* with resources [25]. Therefore, our investigation focus on the following research question.

Can we identify, represent, and interact with heterogeneous streams and events coming from a variety of Web sources?

The research question above implies that streams and events, the key abstractions of Stream Processing, become are valid Web resources. Nevertheless, the nature of streams and events, which as they are respectively *unbounded* and *ephemeral*, contrasts with the nature of Web resources, which are stateful. How this impact identification, representation, and processing is the focus of our investigation. Notably, the seminal work on Stream Reasoning and RDF Stream Processing has paved the road that goes in this direction [21].

To guide the study, we follow the Design Science (DS) research methodology, which studies how *artifacts*, i.e., software components, algorithms, or techniques, interact with a *problem context* that needs improvement. Such interaction, called *treatment*, is designed by researchers to solve a problem. The ultimate goal of DS is to design theories that allow exploring, describing, explaining, and predicting phenomena [51]. The validation of treatments is based on principle compliance, requirements satisfaction, and performance analysis.

Outline. Section 8.2 presents the state-of-the-art on Stream Reasoning and RDF Stream Processing. Section 8.3 formulates the research problems. Section 8.4 presents the major contributions of this research work. Finally, Sect. 8.5 concludes the chapter.

8.2 Background

According to Cugola et al. a data stream is an unbounded sequence of data items ordered by timestamp [14]. On the other hand, an event is an atomic (happened entirely or not at all) domain entity that describes something that happened at a certain time [29]. While streams are infinite and shall be analyzed continuously, events are sudden and ephemeral and shall be detected to react appropriately.

Previous attempts to handle Data Velocity on the Web belong to the Stream Reasoning (SR) [22] and RDF Stream Processing (RSP) [30] literature. Existing works discuss how to identify [7, 39] and represent stream and events [24, 35, 40], but most of the literature focuses on processing RDF streams [11, 19, 20, 30] and detect RDF events [4, 18]. Due to lack of space, we cannot list all the relevant works and we invite the interested readers to consult the recent surveys [21, 30].

Works on *identification* of streams and events directly refer to the notion of Web resource [25]. Barbieri and Della Valle [7] propose to identify streams and each element in the stream. In particular, they propose to use RDF named graphs for both the stream (sGraph) and its elements (iGraphs). The sGraph describes a finite sub-portion the stream Stream made of relevant iGraphs. Moreover, Sequeda and Corcho [39]'s proposal includes a set of URI schemas that incorporate spatio-temporal metadata. This mechanism is suitable to identify sensors and their observations but makes the URI not opaque.

Works on *representation* focus on RDF Streams, i.e., data streams whose data items are timestamped RDF graphs or triples. The Semantic Web community proposed ontological models for modelling such items from a historical point of view. However, neither of these focuses on the infinite nature of streams nor the ephemeral nature of events. Only Schema.org recently included a class that can identify streams as resources.[2] In the context of events, relevant vocabularies are the Linking Open Descriptions of Events (LODE) [40] and Simple Event Model [24]. Efforts on representing RDF streaming data include, but are not limited to FrAPPe [5], which uses pixels as an element for modelling spatio-temporal data, SSN [13] which models sensors observations, and SIOC [33] which include social media micro-posts.

Works on *interaction* divide into protocols for *access* and solutions for *processing* streams and events. The former successfully relies on HTTP extensions for continuous and reactive consumption of Web data, e.g., WebSockets and Server-Sent Events. The latter includes, but it is not limited to works on (i) *Analysis* of streams, i.e., filtering, joining, and aggregating; (ii) *Detection* of events, i.e., matching trigger conditions on the input flow and take action, and (iii) *Reasoning* over and about time, i.e., deducing implicit information from the input streams using inference algorithms that abstract and compose stream elements.

RSEP-QL is the most prominent work in the context of Web stream and event processing [18, 20]. Dell'Aglio et al. build on the state-of-the-art SPARQL extensions for stream analysis event detection. RDF Streams are the data model of choice. Moreover, a continuous query model allows expressing window operations and event patterns over RDF Streams.

Works on reasoning include ontology-based streaming data access (OBSDA) [11], and incremental reasoning [19]. These approaches aim at boosting reasoning performance to meet velocity requirements. Works on OBSDA use query-rewriting to process Web streaming data directly at the source while works on incremental reasoning focus on making reasoning scalable in the presence of frequent updates.

8.3 Problem Statement

According to Design Science, research problems divide into *Design Problems* and *Knowledge Questions*. The former are the problems to (re-)design an artifact so that it

---

[2]https://schema.org/DataFeed.

better contributes to achieving some goal. A solution to a design problem is a design, i.e., a decision about what to do that is generally not unique. The latter include explanatory or descriptive questions about the world. Knowledge questions do not call for a change in the world, and the answer is a unique falsifiable proposition.

From the related literature, it emerges that the *identification* of streams and events using URIs is possible [6, 34, 39]. Moreover, new protocols like WebSockets enable continuous data *access* on the Web. Therefore, our investigation focuses on how to *represent* and *process* streams and events on the Web.

The *representation problem* calls for *improving the Web of Data by enabling conceptual modeling and description of new kinds of resources, i.e., infinite ones like streams and ephemeral ones like events.*

A possible solution to solving the *representation problem* is an ontology, i.e., the specialization of a conceptualizations [23]. However, existing ontologies do not satisfy the requirements of Web users, when they are interested in representing infinite and ephemeral knowledge (cf Sect. 8.2). Existing ontologies were designed to model time-series or historical events that one can query by looking at the past. Moreover, these works neglect the peculiar natures of stream/event transformations, which are continuous, i.e., they last forever [26]. In practice, a shared vocabulary to describe streams and events on the Web is still missing.

The *processing problem* calls for *improving the Web of Data by enabling expressive yet efficient analysis of Web streams and detection of Web events.*

A possible solution to the *processing problem* is to combine semantic technologies and stream processing ones. However, existing solutions show high-performance but limited expressiveness or vice versa, they are very expressive but not efficient [3]. In practice, an expressive yet efficient Stream Reasoning approach that combines existing ones was never realized [42].

Query languages, algorithms, and architectures designed to address the processing problem are typically evaluated using benchmarks like CityBench [2]. However, recent works indicate the lack of a systematic and comparative approach to artifact validation [17, 37]. These limitations focus on two formal properties of the experimental results, i.e., repeatability and reproducibility. The former refers to variations on repeated measurements on the object of study under identical conditions. The latter refers to measurement variations on the object of study under changing experimental conditions.

Therefore, we formulate an additional *validation problem* that calls for *improving validation research by enabling a systematic comparative exploration of the solution space.* A possible solution to the *validation problem* is a methodology that guides researchers to design reproducible and repeatable experiments.

8.4 Major Results

In this section, we present the significant results of our research work. The various contributions are organized according to the identified problems.

Our primary contribution to solving the *representation problem* is the Vocabulary for Cataloging Linked Streams (VoCaLS) [50]. VoCaLS is an OWL 2 QL ontology that includes three modules: a *core* module that enables identifying streams as resources; *service description* that allows describing stream producers and consumers as well as catalogs, and *provenance* that allows auditing continuous transformations. Following the design science methodology, we inquired the community to collect our requirements [38]. Then, to verify VoCaLS compliance, we used the vocabulary in real-world scenarios. Moreover, we validated VoCaLS by expert opinion (peer review) and using Tom Gruber's ontology-design principles [23]. Listing 1 shows an example of a stream description and publication. It shows that VoCaLS allows (i) identifying the stream as a resource, e.g., an RDF Stream; (ii) it provides providing a static stream description including metadata like the license; it enables (iii) accessing the stream content via endpoints that decouple identification (which happens via HTTP) from consumption that uses more appropriate protocols, e.g., WebSockets.

```
PREFIX : <http://linkeddata.stream/resource/> .
<http://linkeddata.stream>  a vsd:PublishingService .
:descriptor  a vocals:streamDescriptor ;
             dcat:dataset :stream1 ;
             dcat:publisher <http://linkeddata.stream>;
             dcat:license <https://creativecommons.org/licenses
                  /by-nc/4.0/> ;
             dcat:description "stream of sensors observations".
:stream1 a vocals:RDFStream ;
    vocals:hasEndpoint [   a vocals:streamEndpoint ;
                           dcat:format frmt:JSON-LD ;
                           dcat:accessURL "ws://example.org/iot
                                /sensor". ] .
```

Listing 1 Example of RDF stream using VoCaLS.

Our primary contribution to solve the *processing problem* is the Expressive Layered Fire-hose (ELF) (formerly streaming MASSIF) [9]. ELF is a stream reasoning platform designed after a renovated Cascading Stream Reasoning (cf Fig. 8.1). ELF can identify the best trade-off between efficiency and expressiveness. To this extent, it organizes the stream reasoners in a layered network and orchestrates the processing to be inversely proportional to the input streams rate.

In the *Continuous Processing* layer (L1), data are in the form of structured streams. L1's operations are elementary and can sustain very high-rates (millions of items per minute). Examples of operations include filters, left-joins, and simple aggregations. In this level, data streams can be converted to foster data integration in the lay-

Fig. 8.1 A renovated
Cascading Reasoning vision
w.r.t. [41]

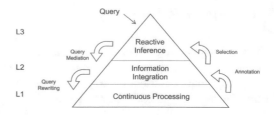

ers above. Possible implementations leverage on window-based stream processing languages for structured data like Streaming SQL [49].

The *Information Integration* layer (L2) aims at building a uniform view over the input streams using a conceptual model. In L2, data streams are typically semi-structured, e.g., RDF Stream. L2's operations are slightly more complex than L1's, e.g., pattern matching over graph streams, and the input rate is reduced to hundreds of thousands items per minute. Moreover, at this level, data streams can be *interpreted* using background domain knowledge. In these regards, our further contribution is C-SPRITE, i.e., an algorithm for efficient hierarchical reasoning over semantic streams [10]. C-SPRITE applies a hybrid reasoning technique that outperforms existing reasoners for Instance Retrieval, even when the number of sub-classes to check is more than a hundred. Possible implementations of this layer include RSP engines, which allows enriching and joining multiple streams, or approaches for Ontology-Based streaming data access [12].

The *Reactive Inference* layer (L3) calls for reactive operations that combines and compare high-level abstractions from various domain. In L3, streams are usually symbolic, e.g., event types, and operations can be very expressive because they deal with a reduced input rated (thousands of items per minute). Possible reasoning framework that are suitable for L3 are Description Logic (DL), temporal logical, and Answer Set Programming (ASP). Our further contribution concerning L3 is Ontology-Based Event RecognitiON (OBERON) (formerly OBEP) [9, 10, 43], i.e., (i) is an A Domain-Specific Language that treats events as first-class objects [43]. OBERON uses two forms of reasoning to detect and compose events over Web streams, i.e., Description Logics reasoning and Complex Event Recognition. Notably, machine learning techniques such as Bayesian networks or hidden Markov models are also suitable approaches for this layer.

The investigation related the *validation problem* develops in [46–48]. Validation research is comparative, and thus, it relies on the notion of experiment [28]. To guarantee repeatability and reproducibility, researchers must have full control over both the experimental environment and the object of study. Thus, our main research contribution to solve the validation problem are (i) a methodology for experiment design for RSP [42] and the architecture for an experimental environment based on the notion of Test-Stand [46], and (ii) a Web environment for experimentation called RSPLab, which guarantees reproducibility and repeatability of experimental results using containerization techniques in the context of RDF Stream Processing.

Moreover, in [44], we highlight the issues related to designing a query language based on RSP-QL formalization that treats streams as first-class objects and keeps the constructs minimal, homogeneous, symmetric and orthogonal [15]. The work evolved into a reference implementation for RSP-QL called YASPER [45] and a framework for rapid-prototyping.[3]

8.5 Conclusion

In this chapter, we summarize the work presented by the research work on representing and processing streams and events on the Web.

With the growing popularity of data catalogs like Google's Dataset Search [8], the research around VoCaLS is potentially impactful. Streams and events are novel kinds of Web resources that are relevant for a number of applications. VoCaLS is a first step towards modelling unbounded and/or ephemeral knowledge. Nevertheless, more work is left to be done in terms of knowledge representation and reasoning. To this extent, an updated version of VoCaLS, which includes better conceptualizations for Web streams and events, is in progress.

Moreover, with the spread of Knowledge Graphs (KG), efficient yet expressive reasoning techniques are relevant as never before. Indeed, KGs are vast and constantly evolving. Therefore, scalable and event-driven reasoning techniques look promising. In particular, the work on C-SPRITE hits a significant trade-off between expressiveness and efficiency. In these regards, pushing efficient reasoning further to incorporate more sophisticated language features, e.g., transitive property, is extremely appealing.

References

1. Alahakoon D, Yu X (2016) Smart electricity meter data intelligence for future energy systems: a survey. IEEE Trans Ind Inform 12(1):425–436
2. Ali MI, Gao F, Mileo A (2015) Citybench: a configurable benchmark to evaluate RSP engines using smart city datasets. In: The Semantic web - ISWC 2015 - 14th international semantic web conference, Bethlehem, PA, USA, 11–15 Oct 2015, proceedings, part II, pp 374–389. https://doi.org/10.1007/978-3-319-25010-6_25
3. Anicic D (2012) Event processing and stream reasoning with ETALIS. Ph.D. thesis, Karlsruhe Institute of Technology
4. Anicic D, Fodor P, Rudolph S, Stojanovic N (2011) EP-SPARQL: a unified language for event processing and stream reasoning. In: WWW. ACM, pp 635–644
5. Balduini M, Della Valle E (2015) Frappe: a vocabulary to represent heterogeneous spatio-temporal data to support visual analytics. In: International semantic web conference (2). Lecture notes in computer science, vol 9367. Springer, Berlin, pp 321–328
6. Barbieri DF, Braga D, Ceri S, Della Valle E, Grossniklaus M (2010) Incremental reasoning on streams and rich background knowledge. In: ESWC (1). Lecture notes in computer science, vol 6088. Springer, Berlin, pp 1–15

---

[3] https://github.com/riccardotommasini/yasper.

7. Barbieri DF, Della Valle E (2010) A proposal for publishing data streams as linked data - A position paper. In: LDOW. CEUR workshop proceedings, vol 628. CEUR-WS.org
8. Benjelloun O, Chen S, Noy NF (2020) Google dataset search by the numbers. https://arxiv.org/abs/2006.06894
9. Bonte P, Tommasini R, Della Valle E, Turck FD, Ongenae F (2018) Streaming MASSIF: cascading reasoning for efficient processing of iot data streams. Sensors 18(11)
10. Bonte P, Tommasini R, Turck FD, Ongenae F, Della Valle E (2019) C-sprite: efficient hierarchical reasoning for rapid RDF stream processing. In: DEBS. ACM, pp 103–114
11. Calbimonte J, Corcho Ó (2014) Evaluating SPARQL queries over linked data streams. Linked data management. Chapman and Hall/CRC, Boca Raton
12. Calbimonte J, Mora J, Corcho Ó (2016) Query rewriting in RDF stream processing. In: ESWC. Lecture notes in computer science, vol 9678. Springer, Berlin, pp 486–502
13. Compton M, Barnaghi PM, Bermudez L, Garcia-Castro R, Corcho Ó, Cox SJD, Graybeal J, Hauswirth M, Henson CA, Herzog A, Huang VA, Janowicz K, Kelsey WD, Phuoc DL, Lefort L, Leggieri M, Neuhaus H, Nikolov A, Page KR, Passant A, Sheth AP, Taylor K (2012) The SSN ontology of the W3C semantic sensor network incubator group. J Web Sem 17:25–32
14. Cugola G, Margara A (2010) TESLA: a formally defined event specification language. In: DEBS. ACM, pp 50–61
15. Date CJ (1984) Some principles of good language design (with especial reference to the design of database languages). SIGMOD Rec 14(3):1–7
16. Della Valle E, Ceri S, van Harmelen F, Fensel D (2009) It's a streaming world! reasoning upon rapidly changing information. IEEE Intell Syst 24(6):83–89
17. Dell'Aglio D, Calbimonte J, Balduini M, Corcho Ó, Della Valle, E (2013) On correctness in RDF stream processor benchmarking. In: International semantic web conference (2). Lecture notes in computer science, vol 8219. Springer, Berlin, pp 326–342
18. Dell'Aglio D, Dao-Tran M, Calbimonte J, Phuoc DL, Della Valle E (2016) A query model to capture event pattern matching in RDF stream processing query languages. In: EKAW. Lecture notes in computer science, vol 10024, pp 145–162
19. Dell'Aglio D, Della Valle E (2014) Incremental reasoning on RDF streams. Linked data management. Chapman and Hall/CRC, Boca Raton, pp 413–435
20. Dell'Aglio D, Della Valle E, Calbimonte J, Corcho Ó (2014) RSP-QL semantics: a unifying query model to explain heterogeneity of RDF stream processing systems. Int J Semantic Web Inf Syst 10(4)
21. Dell'Aglio D, Della Valle E, van Harmelen F, Bernstein A (2017) Stream reasoning: a survey and outlook. Data Sci 1(1–2):59–83
22. Germano S, Pham T, Mileo A (2015) Web stream reasoning in practice: on the expressivity vs. scalability tradeoff. In: RR. Lecture notes in computer science, vol 9209. Springer, Berlin, pp 105–112
23. Gruber TR (1995) Toward principles for the design of ontologies used for knowledge sharing? Int J Hum-Comput Stud 43(5–6):907–928
24. van Hage WR, Malaisé V, Segers R, Hollink L, Schreiber G (2011) Design and use of the simple event model (SEM). J Web Sem 9(2):128–136
25. Jacobs I, Walsh N (2004) Architecture of the world wide web, volume one. W3C recommendation, W3C. https://www.w3.org/TR/webarch/
26. Keskisärkkä R (2016) Representing RDF stream processing queries in RSP-SPIN. In: International semantic web conference (posters & demos). CEUR workshop proceedings, vol 1690. CEUR-WS.org
27. Kolozali S, Bermúdez-Edo M, Puschmann D, Ganz F, Barnaghi PM (2014) A knowledge-based approach for real-time iot data stream annotation and processing. In: 2014 IEEE international conference on internet of things, Taipei, Taiwan, 1–3 Sep 2014, pp 215–222
28. Kuehl RO (2000) Design of experiments stastistical principles of research design and analysis. No. Q182. K84 2000
29. Luckham D (2008) The power of events: an introduction to complex event processing in distributed enterprise systems. In: RuleML. Lecture notes in computer science, vol 5321. Springer, Berlin, p 3

30. Margara A, Urbani J, van Harmelen F, Bal HE (2014) Streaming the web: reasoning over dynamic data. J Web Sem 25:24–44
31. Mathioudakis M, Koudas N (2010) Twittermonitor: trend detection over the twitter stream. In: SIGMOD conference. ACM, pp 1155–1158
32. Nixon LJB, Fischl D, Scharl A (2019) Real-time story detection and video retrieval from social media streams. In: Mezaris V, Nixon LJB, Papadopoulos S, Teyssou D (eds.) Video verification in the fake news era. Springer, Berlin, pp 17–52. https://doi.org/10.1007/978-3-030-26752-0_2
33. Passant A, Bojars U, Breslin JG, Decker S (2009) The SIOC project: semantically-interlinked online communities, from humans to machines. In: COIN@AAMAS&IJCAI&MALLOW. Lecture notes in computer science, vol 6069. Springer, Berlin, pp 179–194
34. Pimentel V, Nickerson BG (2012) Communicating and displaying real-time data with web-socket. IEEE Internet Comput 16(4):45–53
35. Rinne M, Blomqvist E, Keskisärkkä R, Nuutila E (2013) Event processing in RDF. In: WOP. CEUR workshop proceedings, vol 1188
36. Russom P, et al (2011) Big data analytics. TDWI best practices report, fourth quarter
37. Scharrenbach T, Urbani J, Margara A, Della Valle E, Bernstein A (2013) Seven commandments for benchmarking semantic flow processing systems. In: The semantic web: semantics and big data, 10th international conference, ESWC 2013, Montpellier, France, 26–30 May 2013. Proceedings, pp 305–319. https://doi.org/10.1007/978-3-642-38288-8_21
38. Sedira YA, Tommasini R, Della Valle E (2017) Towards vois: a vocabulary of interlinked streams. In: DeSemWeb. CEUR workshop proceedings, vol 1934. CEUR-WS.org
39. Sequeda JF, Corcho Ó (2009) Linked stream data: a position paper. In: SSN. CEUR workshop proceedings, vol 522, pp 148–157. CEUR-WS.org
40. Shaw R, Troncy R, Hardman L (2009) LODE: linking open descriptions of events. In: ASWC. Lecture notes in computer science, vol 5926. Springer, Berlin, pp 153–167
41. Stuckenschmidt H, Ceri S, Della Valle E, van Harmelen F (2010) Towards expressive stream reasoning. In: Semantic challenges in sensor networks. Dagstuhl seminar proceedings, vol 10042. Schloss Dagstuhl - Leibniz-Zentrum für Informatik, Germany
42. Tommasini R (2015) Efficient and expressive stream reasoning with object-oriented complex event processing. In: DC@ISWC. CEUR workshop proceedings, vol 1491. CEUR-WS.org
43. Tommasini R, Bonte P, Della Valle E, Ongenae F, Turck FD (2018) A query model for ontology-based event processing over RDF streams. In: EKAW. Lecture notes in computer science, vol 11313. Springer, Berlin
44. Tommasini R, Della Valle E (2017) Challenges & opportunities of RSP-QL implementations. In: WSP/WOMoCoE. CEUR workshop proceedings, vol 1936, pp 48–57. CEUR-WS.org
45. Tommasini R, Della Valle E (2017) Yasper 1.0: towards an RSP-QL engine. In: Proceedings of the ISWC 2017 posters & demonstrations and industry tracks co-located with 16th international semantic web conference (ISWC)
46. Tommasini R, Della Valle E, Balduini M, Dell'Aglio D (2016) Heaven: a framework for systematic comparative research approach for RSP engines. In: ESWC. Lecture notes in computer science, vol 9678. Springer, Berlin, pp 250–265
47. Tommasini R, Della Valle E, Balduini M, Sakr S (2020) On teaching web stream processing - lessons learned. In: Bieliková M, Mikkonen T, Pautasso C (eds.) Web engineering - 20th international conference, ICWE 2020, Helsinki, Finland, 9–12 June 2020, proceedings. Lecture notes in computer science, vol 12128. Springer, Berlin, pp 485–493. https://doi.org/10.1007/978-3-030-50578-3_33
48. Tommasini R, Della Valle E, Mauri A, Brambilla M (2017) Rsplab: RDF stream processing benchmarking made easy. In: ISWC, pp 202–209
49. Tommasini R, Sakr S, Valle ED, Jafarpour H (2020) Declarative languages for big streaming data. In: Bonifati A, Zhou Y, Salles MAV, Böhm A, Olteanu D, Fletcher GHL, Khan A, Yang B (eds.) Proceedings of the 23nd international conference on extending database technology, EDBT 2020, Copenhagen, Denmark, March 30 - April 02, 2020. pp. 643–646. OpenProceedings.org. https://doi.org/10.5441/002/edbt.2020.84

50. Tommasini R, Sedira YA, Dell'Aglio D, Balduini M, Ali MI, Phuoc DL, Della Valle E, Cal-
    bimonte J (2018) Vocals: Vocabulary and catalog of linked streams. In: International semantic
    web conference (2). Lecture notes in computer science, vol 11137. Springer, Berlin, pp 256–272
51. Wieringa R (2014) Design science methodology for information systems and software engi-
    neering. Springer, Berlin

Open Access This chapter is licensed under the terms of the Creative Commons Attribution 4.0
International License (http://creativecommons.org/licenses/by/4.0/), which permits use, sharing,
adaptation, distribution and reproduction in any medium or format, as long as you give appropriate
credit to the original author(s) and the source, provide a link to the Creative Commons license and
indicate if changes were made.

The images or other third party material in this chapter are included in the chapter's Creative
Commons license, unless indicated otherwise in a credit line to the material. If material is not
included in the chapter's Creative Commons license and your intended use is not permitted by
statutory regulation or exceeds the permitted use, you will need to obtain permission directly from
the copyright holder.

Chapter 9
Preplay Communication in Multi-Player Sequential Games: An Overview of Recent Results

Andrea Celli

9.1 Introduction

The computational study of game-theoretic solution concepts is fundamental to describe the optimal behavior of rational agents interacting in a strategic setting, and to predict the most likely outcome of a game. Equilibrium computation techniques have been applied to numerous real-world problems. Among other applications, they are the key building block of the best poker-playing AI agents [5, 6, 27], and have been applied to physical and cybersecurity problems (see, e.g., [18, 20, 21, 30–32]).

In this section, we start by presenting a simple example and reviewing the fundamental model of sequential games with imperfect information. Then, we give a brief overview of problems involving preplay communication, which will be the focus of this summary.

9.1.1 Motivating Example

The following simple example shows that, even in toy problems, preplay communication can give a significant advantage to players that decide to implement it.

Example 1 (A simple coordination game) There are three players and each player has two available actions (which are denoted by LEFT and RIGHT). Moreover, player 1 and player 2 have identical goals, while player 3 has the opposite objectives. Therefore, player 1 and 2 will try to maximize their expected utilities and player 3 will act as an adversary, whose goal is minimizing player 1 and 2's expected utility.

A. Celli (✉)
Politecnico di Milano, Milano, Italy
e-mail: andrea.celli@polimi.it

© The Author(s) 2021
A. Geraci (ed.), *Special Topics in Information Technology*,
PoliMI SpringerBriefs, https://doi.org/10.1007/978-3-030-62476-7_9

The strategic interaction goes as follows: as the game starts, each player has to select one action, without observing the choice of the others. Player 1 and 2 receive a payoff of $K > 0$ if they correctly guess the action taken by player 3 (e.g., when player 3 plays LEFT, player 1 and 2 are rewarded K if they both play LEFT), and they receive payoff 0 otherwise. Player 3 receives a payoff of $-K$ whenever player 1 and 2 receive payoff K, and 0 otherwise.

The best that player 3 can do in order to fool the other players is selecting an action between LEFT/RIGHT with equal probability. Intuitively, this is the strategy of player 3 which makes it more difficult for the others to guess the action he/she chose. On the other hand, since player 1 and 2 have equal objectives, it may be profitable for them to jointly plan their strategies before the beginning of the game. Specifically, they could decide to play actions (LEFT, LEFT) with probability 0.5, and actions (RIGHT, RIGHT) with probability 0.5. In this way, they would be able to avoid outcomes with rewards always equal to 0 in which they play different actions from each other. Before the beginning of the game, player 1 and 2 could toss a coin and select whether to play (LEFT, LEFT) or (RIGHT, RIGHT) depending on the outcome of the coin toss. In this way, the expected utility of player 1 and 2 would be $K/2$ as they would reach each outcome with payoff K with probability 0.25. If player 1 and 2 decided not to communicate before the beginning of the game the best that they could do is employing player 3's strategy (i.e., selecting one of their actions randomly). In this way, player 1 and player 2 would get a payoff of $K/4$, since each outcome with payoff K is reached with probability 0.125.

Even in such a simple game the possibility of exploiting preplay communication yield an increase of 50% in the expected utility of players who adopted it. This increase can be arbitrarily large in game instances which are slightly more complex [13]. In the remainder of this summary, we will discuss how to model such strategic scenarios and review some recent results related to these problems.

9.1.2 Sequential Games with Imperfect Information

We focus on a powerful model of strategic interaction which can model sequential moves, imperfect information, and outcome uncertainty. An *extensive-form game* (EFG) (*a.k.a.* a *sequential game*) models a sequential interaction among players. An EFG is represented as a game tree, where each node is identified by the ordered sequence of actions leading to it from the root node. Each node represents a decision point of the game and is associated to a single player, who has a set of available actions at that node represented by its branches. A payoff for each player is associated to each leaf node (*terminal node*) of the game tree. Finally, exogenous stochasticity is modeled via a virtual player (*a.k.a.* nature or *chance*) that plays non-strategically (i.e., it plays according to a fixed strategy). Chance is used to describe, e.g., the probability of receiving a certain hand in a card playing game. We say that a two-player game is *zero-sum* if, for each terminal node, the sum of the utilities of the

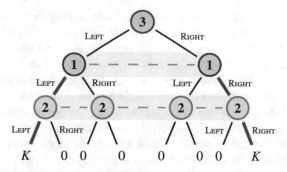

Fig. 9.1 The game described in Example 1. Decision nodes of each player are denoted by the number of the player taking action at that decision point. Each player has a single information set. Nodes belonging to the same information set are within the same grey area. Terminal nodes show the payoffs of player 1 and 2. Branches of the tree in red highlight the pairs of actions that player 1 and 2 should play when using preplay communication

two players equals 0. A game is *constant-sum* if, for each terminal node, the sum of players' utilities is equal to a certain constant. If this does not hold, we say that the game is *general-sum*.

In general, a player may not be able to observe all the other players' actions, and players may have information on the state of the game which is not shared (i.e., in poker each player does not know other players' hands). Imperfect information is represented via *information sets* (or *infosets*), which group together decision nodes of a certain player that are indistinguishable to her. We assume players have *perfect recall*, that is they have perfect memory of their past actions and observations. Figure 9.1 shows the EFG representing the game described in Example 1.

The most widely adopted notion of equilibrium is that of *Nash equilibrium* [29]: each player should not have incentives in deviating from his/her strategy, assuming the other players do not deviate either. In Example 1, players reach a Nash equilibrium when each of them selects an action according to a uniform probability distribution over the available actions.

9.1.3 Preplay Communication

In the context of imperfect-information games, a vast body of literature focuses on the computation of Nash equilibria in two-player, zero-sum games (see, e.g., [19, 33, 34]), where recent results demonstrated that it is possible to compute strong solutions in theory and practice.

While relevant, two-player, zero-sum games are rather restrictive, as many practical scenarios are not zero-sum and involve more than two players. Moreover, especially in general-sum games, the adoption of a Nash equilibrium may present some difficulties when used as a prescriptive tool. Indeed, when multiple Nash equilibria coexist, the model prevents players from synchronizing their strategies, since communication between players is prohibited. In real-world scenarios, where some form of communication among players is usually possible, different solution concepts are required as communication allows players for coordinated behaviors.

We focus on scenarios where players can exploit *preplay communication* [24, 25], i.e., players have an opportunity to discuss and agree on tactics before the game starts, but will be unable to communicate during the game. Consider, as an illustration, the case of a poker game where multiple players are colluding against an identified target player. Colluders can agree on shared tactics before the beginning of the game, but are not allowed any explicit communication while playing. In other settings, players might be forced to cooperate by the nature of the interaction itself. This is the case, for instance, in Bridge. Preplay players' coordination introduces new challenges with respect to the case in which agents take decisions individually, as understanding how to coordinate before the beginning of the game requires reasoning over the entire game tree. It is easy to see that this causes an exponential blowup in the agents' action space and, therefore, even relatively small game instances are usually deemed intractable in this setting.

When modeling preplay communication, it is instructive to introduce an additional agent, called the *mediator*, that does not take part in the game, but may send signals (usually actions' recommendations) to other players just before the beginning of the game. In the following sections, we explore different forms of preplay coordination in sequential games. The scenarios we consider can be classified through the following questions: (i) *who is receiving the mediator's recommendations?* (ii) *do players have similar goals?* (iii) *is the mediator self-interested? and does the mediator have more information on the state of the game than other players?* When the mediator is sending signals only to members of the same team we talk about *adversarial team games* (Sect. 9.2). In more general settings, players may not have identical objectives. In these scenarios, the mediator has to accurately plan incentives for each individual player and the problem becomes the computation of a *correlated equilibrium* (Sect. 9.3). Finally, the mediator may hold more information than the players of the game, and he/she may be willing to use this information asymmetry to achieve his/her own goals. This setting is modeled via the Bayesian persuasion framework (Sect. 9.4).

9.2 Adversarial Team Games

A recent line of research focuses on preplay communication in *team games*. A *team* of agents is defined as a set of players sharing the same objectives. Following this simple definition, player 1 and 2 of Example 1 form a team. An interesting problem is understanding how team members can coordinate their actions when facing an opponent (e.g., player 3 in the example). We call these games *adversarial team games*. Even without communication during the game, the planning phase gives the team members an advantage: for instance, the team members could skew their strategies to use certain actions to signal about their state (for example, in card-playing games, the current hands they're holding). In other words, by having agreed on each member's planned reaction under any possible circumstance of the game, information can be silently propagated in the clear, by simply observing public information.

Initially, adversarial team games where studied in games with simultaneous moves [2, 4]. Celli and Gatti [13] first studied the setting in which a team of agents faces an adversary in a sequential interaction. This work formally defines the game model and shows that different forms of intra-team communication result in different models of coordination: (i) a mediator that can send and receive *intraplay* signals (i.e., messages are exchanged during the execution of the game); (ii) a mediator that only exploits preplay communication, sending recommendations just before the beginning of the game; (iii) team members jointly plan their strategies, but have no access to a mediator to synchronize action execution. The main focus has been on the second scenario, where only preplay communication is possible. Scenarios (i) and (iii) are instructive to understand the advantages of different forms of intra-team communication. These different coordination capabilities are compared via the analysis of inefficiency indexes measuring the relative losses in the team's expected utility. Interestingly, their experimental evaluation shows that, in practice, preplay communication is often enough to reach near-optimal performances. An application of these techniques is the work of Basilico et al. [3], where team games are used to coordinate patrollers in environments at risk.

Motivated by the complexity of the problem of coordinating team members with preplay communication, Farina et al. [23] present a scalable learning algorithm to compute an approximate solution to this problem. In doing so, the authors highlight a strong analogy with imperfect-recall games, and propose a new game representation, called *realization form*, which can also be applied to this setting. Then, they exploit the new representation to derive an auxiliary construction that allows one to map the problem of finding an optimal coordinated strategy for the team to the well-understood Nash equilibrium-finding problem in a (larger) two-player zero-sum perfect-recall extensive-form game. By reasoning over the auxiliary game, they devise an anytime algorithm, *fictitious team-play*, that is guaranteed to converge to an optimal coordinated strategy for the team against an optimal opponent. Then, they demonstrate the scalability of the learning algorithm on standard imperfect-information test instances (such as, Leduc hold'em poker and Goofspiel).

9.3 Correlated Equilibria in Sequential Games

The members of a team share the same objectives. Therefore, the mediator does not have to enforce any incentive-compatibility constraint. However, it may happen that agents receiving mediator's recommendations do not share the same objectives. In this case, the mediator has to incentivize each player to follow moves' recommendations. Here, the mediator is assumed to be *benevolent*, i.e., she aims at maximizing the expected social welfare of the game.

We briefly discuss some works investigating whether correlation can be reached efficiently even in settings where players have limited communication capabilities (i.e., they can only observe signals before the beginning of the game). Therefore, we focus on sequential games in which only preplay communication is admitted, and study correlated equilibria that allow the mediator to recommend actions just *before* the playing phase of the game (namely, the *correlated equilibrium* (CE) [1] and the *coarse correlated equilibrium* (CCE) [28]).

These problems have been proved to be computationally hard in most settings. Celli et al. [11] provide several results characterizing the complexity of computing optimal (i.e., social welfare maximizing) CEs and CCEs, and their approximation complexity. First, in an extended version of the paper, they prove that approximating an optimal (i.e., social welfare maximizing) CE is not in Poly-APX even in two player games without chance moves, unless P = NP. Next, they identify the conditions for which finding an optimal CCE is NP-hard. However, they show that an optimal CCE can be found in polynomial-time in two-player extensive-form games without chance moves. Finally, Celli et al. [14] complete the picture on the computational complexity of finding social-welfare-maximizing CCEs by showing that the problem is not in Poly-APX, unless P = NP, in games with three or more players (chance included).

There are various algorithms for computing CCEs in general-sum, multi-player, sequential games. Celli et al. [11] provide a column generation framework to compute optimal CCEs in practice, and show how to generalize it to the *hard cases* of the problem. Celli et al. [14] focus on the problem of computing *an ε-CCE* (i.e., an approximate CCE). The authors present an enhanced version of CFR [34] which computes an average correlated strategy which is guaranteed to converge to an approximate CCE with a bound on the regret which is sub-linear in the size of the game tree.

9.4 Bayesian Persuasion with Sequential Games

Finally, it may happen that the mediator is self-interested, and may exploit asymmetries in the availability of information to design a signaling scheme, in order to persuade players to select favorable actions. In this setting, the mediator is looking for a way to coordinate the individual behavior of each player in order to reach a preferred outcome of the game.

Celli et al. [12] examine information-structure design problems as a means of forcing coordination towards a certain objective. More precisely, they start from the usual scenario where a mediator can communicate action recommendations to players before the beginning of a sequential game. Suppose that parties (i.e., the mediator and the players) are asymmetrically informed about the current state of the game. Specifically, the mediator is able to observe more information than the other players. Celli et al. [12] pose the following question: *can the mediator exploit the information asymmetry to coordinate players' behavior toward a favorable outcome?*

This problem can be accurately modeled via the *Bayesian persuasion* framework [26]. Celli et al. [12] investigate private persuasion problems with multiple receivers interacting in a sequential game, and study the continuous optimization problem of computing a private signaling scheme which maximizes the sender's expected utility. The authors show how to address sequential, multi-receiver settings algorithmically via the notion of *ex ante* persuasive signaling scheme, where the receivers commit to following the sender's recommendations having observed only the signaling scheme. They show that an optimal *ex ante* signaling scheme may be computed in polynomial time in settings with two receivers and independent action types, which makes *ex ante* persuasive signaling schemes a persuasion tool which is applicable in practice. Moreover, they show that this result cannot be extended to settings with more than two receivers, as the problem of computing an optimal *ex ante* signaling scheme becomes NP-hard.

9.5 Discussion and Future Research

The research on equilibrium computation in general-sum, multi-player, sequential games has not yet reached the level of maturity reached in the two-player, zero-sum setting, where it is possible to compute strong solutions in theory and practice. In these settings, equilibrium selection problems may render the choice of the appropriate solution concept not obvious, since the Nash equilibrium may not be the appropriate one. Many practical scenarios allow for some form of communication, mitigating the equilibrium selection issue. In this summary, we presented some multi-player problems where players can reach some form of coordination via preplay communication.

There are many interesting questions that need to be addressed in the future. We outline some of them for each of the settings we described. First, it would be interesting to develop a scalable end-to-end approach to learning an optimal team coordinated strategy without prior domain knowledge. Some works going in this direction are Chen et al. [17], Celli et al. [10]. Moreover, as pointed out by Celli et al. [16], the study of algorithms for team games could shed further light on how to deal with imperfect-recall games, which are receiving increasing attention in the community due to the application of imperfect-recall abstractions to the computation of strategies for large sequential games. As for the computation of correlated equilibria in sequential games, it would be interesting to further investigate whether it is possible to define regret-minimizing procedures for general EFGs leading to refinements of the CCEs, such as EFCCEs [22]. A recent work studying a related problem is Celli et al. [15]. Finally, it would be interesting to complement recent works on Bayesian persuasion problems such as Castiglioni et al. [7–9] with scalable algorithms that can be effectively applied to real-world problems.

References

1. Aumann R (1974) Subjectivity and correlation in randomized strategies. J Math Econ 1(1):67–96
2. Basilico N, Celli A, De Nittis G, Gatti N (2017) Team-maxmin equilibrium: efficiency bounds and algorithms. In: AAAI conference on artificial intelligence (AAAI)
3. Basilico N, Celli A, De Nittis G, Gatti N (2017) Coordinating multiple defensive resources in patrolling games with alarm systems. In: Proceedings of the 16th conference on autonomous agents and multiagent systems (AAMAS), pp 678–686
4. Basilico N, Celli A, De Nittis G, Gatti N (2017) Computing the team–maxmin equilibrium in single–team single–adversary team games. Intell Artif
5. Brown N, Sandholm T (2017) Superhuman AI for heads-up no-limit poker: libratus beats top professionals. Science eaao1733:2017
6. Brown N, Sandholm T (2019) Superhuman ai for multiplayer poker. Science 365(6456):885–890
7. Castiglioni M, Celli A, Gatti N (2019) Persuading voters: it's easy to whisper, it's hard to speak loud. arXiv:1908.10620
8. Castiglioni M, Celli A, Gatti N (2020) Public bayesian persuasion: being almost optimal and almost persuasive. arXiv:2002.05156
9. Castiglioni M, Celli A, Marchesi A, Gatti N (2020) Signaling in bayesian network congestion games: the subtle power of symmetry. arXiv:2002.05190
10. Celli A, Ciccone M, Bongo R, Gatti N (2019) Coordination in adversarial sequential team games via multi-agent deep reinforcement learning. arXiv:1912.07712
11. Celli A, Coniglio S, Gatti N (2019) Computing optimal ex ante correlated equilibria in two-player sequential games. In: AAMAS
12. Celli A, Coniglio S, Gatti N (2020) Private bayesian persuasion with sequential games. In: AAAI
13. Celli A, Gatti N (2018) Computational results for extensive-form adversarial team games. In: AAAI
14. Celli A, Marchesi A, Bianchi T, Gatti N (2019) Learning to correlate in multi-player general-sum sequential games. In: NeurIPS
15. Celli A, Marchesi A, Farina G, Gatti N (2020) No-regret learning dynamics for extensive-form correlated and coarse correlated equilibria. arXiv:2004.00603
16. Celli A, Romano G, Gatti N (2019) Personality-based representations of imperfect-recall games. In: Proceedings of the 18th international conference on autonomous agents and multi-agent systems, AAMAS, pp 1868–1870
17. Chen L, Guo H, Du Y, Fang F, Zhang H, Zhu Y, Zhou M, Zhang W, Wang Q, Yu Y (2019) Signal instructed coordination in cooperative multi-agent reinforcement learning
18. Chen X, Han Z, Zhang H, Xue G, Xiao Y, Bennis M (2017) Wireless resource scheduling in virtualized radio access networks using stochastic learning. IEEE Trans Mobile Comput 17(4):961–974
19. Daskalakis C, Deckelbaum A, Kim A (2015) Near-optimal no-regret algorithms for zero-sum games. Games Econ Behav 92:327–348
20. Fang F, Nguyen TH, Pickles R, Lam WY, Clements GR, An B, Singh A, Tambe M, Lemieux A (2016) Deploying paws: Field optimization of the protection assistant for wildlife security. In: Twenty-eighth IAAI conference
21. Fang F, Stone P, Tambe M (2015) When security games go green: designing defender strategies to prevent poaching and illegal fishing. In: Twenty-fourth international joint conference on artificial intelligence
22. Farina G, Bianchi T, Sandholm T (2019) Coarse correlation in extensive-form games. arXiv:1908.09893
23. Farina G, Celli A, Gatti N, Sandholm T (2018) Ex ante coordination and collusion in zero-sum multi-player extensive-form games. In: NeurIPS

24. Forges F (1986) An approach to communication equilibria. Econometrica 54:1375–1385
25. Forges F (2006) Correlated equilibrium in games with incomplete information revisited. Theory Decision 61(4):329–344
26. Kamenica E, Gentzkow M (2011) Bayesian persuasion. Am Econ Rev 101(6):2590–2615
27. Matej M, Martin S, Neil B, Viliam L, Dustin M, Nolan B, Trevor D, Kevin W, Michael J, Michael B (2017) Deepstack: expert-level artificial intelligence in heads-up no-limit poker. Science 356(6337):aam6960
28. Moulin H, Vial J-P (1978) Strategically zero-sum games: the class of games whose completely mixed equilibria cannot be improved upon. Int J Game Theory 7(3):201–221
29. Nash J (1951) Non-cooperative games. Ann Math 54:286–295
30. Píbil R, Lisý V, Kiekintveld C, Bošanský B, Pěchouček M (2012) Game theoretic model of strategic honeypot selection in computer networks. In: International conference on decision and game theory for security, pp 201–220. Springer, Berlin
31. Schlenker A, Xu H, Guirguis M, Kiekintveld C, Sinha A, Tambe M, Sonya SY, Balderas D, Dunstatter N (2017) Don't bury your head in warnings: a game-theoretic approach for intelligent allocation of cyber-security alerts. In: IJCAI, pp 381–387
32. Milind T (2011) Security and game theory: algorithms, deployed systems, lessons learned. Cambridge University Press, Cambridge
33. von Stengel B (1996) Efficient computation of behavior strategies. Games Econ Behav 14(2):220–246
34. Zinkevich M, Bowling M, Johanson M, Piccione C (2007) Regret minimization in games with incomplete information. In: Proceedings of the annual conference on neural information processing systems (NIPS)

Open Access This chapter is licensed under the terms of the Creative Commons Attribution 4.0 International License (http://creativecommons.org/licenses/by/4.0/), which permits use, sharing, adaptation, distribution and reproduction in any medium or format, as long as you give appropriate credit to the original author(s) and the source, provide a link to the Creative Commons license and indicate if changes were made.

The images or other third party material in this chapter are included in the chapter's Creative Commons license, unless indicated otherwise in a credit line to the material. If material is not included in the chapter's Creative Commons license and your intended use is not permitted by statutory regulation or exceeds the permitted use, you will need to obtain permission directly from the copyright holder.

Part IV
Systems and Control

Chapter 10
Leadership Games: Multiple Followers, Multiple Leaders, and Perfection

Alberto Marchesi

10.1 Introduction

Over the last years, *algorithmic game theory* has received growing interest in AI, as it allows to tackle complex real-world scenarios involving multiple artificial agents engaged in a competitive interaction. These settings call for rational agents endowed with the capability of reasoning strategically, which is achieved by exploiting *equilibrium* concepts from game theory. The challenge is to design scalable computational tools that enable the adoption of such equilibrium notions in real-world problems.

The recent advances in the development of equilibrium-finding techniques have lead to the successful application of game-theoretic models in real-world settings. For instance, game theory has been extensively adopted in security domains, with the goal of devising protection strategies which are robust against strategic attackers [47]. Other application domains are found in the *Internet*, where interactions involving multiple strategic agents naturally arise, given the intrinsic distributed nature of the network. One examples is, among others, the problem of designing auction mechanisms for web advertising [23, 27]. Moreover, great achievements have been made towards the development of artificial agents capable of beating human professional in large two-player zero-sum recreational games like Chess [11], Go [46], and Poker [9, 10].

Despite the great attention devoted to algorithmic game theory in the last years, most of the works in the literature study (relatively) simple settings involving only two players with opposite objectives, i.e., two-player zero-sum games. In such models, there is a clear and well-established definition of solution, in which each player aims to maximize her utility given that the opponent acts so as to minimize it. In zero-sum games, this definition corresponds to that of *Nash equilibrium*. Thus, considerable efforts have been devoted to studying the problem of computing (possibly approximate) Nash equilibria in such settings. Instead, more complex games where

A. Marchesi (✉)
Politecnico di Milano, Piazza Leonardo da Vinci 32, Milan, Italy
e-mail: alberto.marchesi@polimi.it

© The Author(s) 2021
A. Geraci (ed.), *Special Topics in Information Technology*,
PoliMI SpringerBriefs, https://doi.org/10.1007/978-3-030-62476-7_10

there are more than two players and/or arbitrary, i.e., general-sum, utilities are widely unexplored. In such scenarios, there is no clear definition of solution to a game, as this strongly depends on the specific application that one wish to represent. As a result, many solution concepts other than the Nash equilibrium have been introduced and studied. However, there is still a lot of work to be done on the computational side, as the algorithmic works on multi-player general-sum games are only few.

In this work, we study settings beyond two-player zero-sum games, focusing on a particular game paradigm which leads to the definition of what is known in the literature as the *Stackelberg equilibrium*.

10.2 The Stackelberg Paradigm

The Stackelberg paradigm was originally introduced by von Stackelberg in 1934 to model economic situations where a firm (the *leader*) moves first and, then, another firm (the *follower*) moves second by reacting to the first firm's move [50]. Recently, this paradigm was brought to new attention by the work of [7], who study a variant of the original Stackelberg paradigm in which the leader commits to a (possibly randomized, i.e., mixed) strategy beforehand, while the follower decides how to play after observing the leader's strategy. In general settings involving multiple players, a *Stackelberg game* is characterized by a group of players who act as leaders with the ability to commit to (possibly mixed) strategies beforehand, whereas the other players are followers who observe the commitment and decide how to play thereafter.

Over the last years, Stackelberg games and their corresponding Stackelberg equilibria have received growing attention in the AI literature, where the computational problem of finding such equilibria in often referred to as the problem of *computing optimal strategies to commit to* [20]. This surge of interest was motivated by the successful applications of Stackelberg games in many interesting real-world settings. In particular, among the others, the security domain is the most explored one, and, in it, different game models have been introduced, usually referred to as *security games* [2, 29, 42, 47]. In such models, there is a defender that has to protect some valuable targets from an attacker, who can wait while observing the defender's protection strategy before deciding where, when and how to attack. This scenario naturally fits into the Stackelberg model, where the defender is the leader and the attacker is the follower. Other interesting applications are found in *toll-setting games*, where the leader is a central authority which collects tolls from the users of a network who, acting as followers, decide on how to best travel through the network so as to minimize their cost after observing the pricing strategy chosen by the authority [30, 31]. Besides the security domain and toll-setting games, applications of Stackelberg games can be found in, among others, interdiction games [12, 39], network routing [1], inspection games [3], and mechanism design [44].

Despite the attention that Stackelberg games received from the AI literature, most of the works related to them focus, with some exceptions (see, e.g., [7, 19, 26]), on particular game settings that involve only two players (i.e., one leader and one follower) and enjoy specific structures, as it is the case in security games. It is worth pointing out two works that study general Stackelberg games with a single leader and multiple follower; specifically, [7] study the case in which the followers play a Nash equilibrium given the leader's commitment, whereas [19] address the case where they play a *correlated equilibrium*.

Let us also notice that, while some works (see, e.g., [8, 16, 32]) address the computation of Stackelberg equilibria in games with a sequential (i.e., tree-form) structure, none of them investigates refinements of such equilibria. This is surprising as refinements have been extensively studied for the Nash equilibrium, since it is well-known that classical (unrefined) solution concepts may lead to a sub-optimal behavior off the equilibrium path in games with a sequential structure (see [24, 49] for some references on the topic).

10.3 Stackelberg Games with Multiple Followers

We address Stackelberg games with a *single leader* and *multiple followers*. Following [7], we study settings in which, after observing the leader's commitment, the followers play a Nash equilibrium in the resulting game. We refer to this solution as *Stackelberg-Nash equilibrium*. We focus on the case in which the followers are restricted to pure (i.e., non-mixed) strategies, as the general problem with followers playing mixed strategies is already known to be computationally intractable [4]. This restriction leads to interesting computational complexity results. Moreover, this is without loss of generality in games that always admit pure-strategy Nash equilibria, as it is the case for congestion games [43].

We study the problem of computing Stackelberg-Nash equilibria, focusing on two cases: the one in which the followers break ties in favor of the leader (what is usually referred to as a *strong* equilibrium), and the case where they break ties against the leader (leading to a *weak* equilibrium). Moreover, we analyze three different classes of games, namely, *normal-form* games, *polymatrix* games, and *congestion* games.

Table 10.1 shows our contributions related to Stackelberg games with a single leader, summarizing the computational complexity and the algorithmic aspects of the problems we study, with focus on normal-form, extensive-form, Bayesian, and poly-matrix games. The table also shows, for comparison, other state-of-the-art results, including those about single-leader single-follower Stackelberg games (our original contributions are those without a reference). Our contributions on Stackelberg congestion games are instead detailed in Table 10.2.

10.3.1 Norma-Form Stackelberg Games

Since a strong Stackelberg-Nash equilibrium (with followers restricted to pure strategies) can be computed efficiently (in polynomial time) by solving multiple *linear programs* (LPs), we entirely devote our analysis to the weak case (with, again, followers restricted to pure strategies). In terms of computational complexity, we show that, differently from the strong case, in the weak one the equilibrium-finding prob-

Table 10.1 Summary of the results on the computation of Stackelberg equilibria in normal-form Stackelberg games, Bayesian Stackelberg games, extensive-form Stackelberg games, and Stackelberg polymatrix games. The state-of-the-art results are those with related references

		Strong Stackelberg(-Nash) equilibrium		Weak Stackelberg(-Nash) equilibrium	
Followers' strategies		Pure	Mixed	Pure	Mixed
Normal-form Stackelberg games					
$n = 2$	Complexity	P [20]		P [7]	
	Algorithm	Multi-LP [20]		Multi-LP [7]	
$n = 3$	Complexity	P	NP-hard, $\notin$ Poly-APX [6]	NP-hard	NP-hard, $\notin$ Poly-APX [6]
	Algorithm	Multi-LP	Spatial branch-and-bound [5]	Multi-lex-MILP	–
$n \geq 4$	Complexity	P	NP-hard, $\notin$ Poly-APX [6]	NP-hard, $\notin$ Poly-APX	NP-hard, $\notin$ Poly-APX [6]
	Algorithm	Multi-LP	Spatial branch-and-bound [5]	Multi-lex-MILP	–
Bayesian Stackelberg games					
$n = 2$	Complexity	NP-hard [20], Poly-APX-complete [33]		NP-hard, Poly-APX-complete	
	Algorithm	MILP [42]		Multi-LP	
Extensive-form Stackelberg games					
$n = 2$	Complexity	NP-hard [32]		NP-hard	
	Algorithm	MILP [8, 16]		Multi-LP [7]	
Stackelberg polymatrix games					
$n = 3$	Complexity	P	NP-hard, $\notin$ Poly-APX [6]	NP-hard	NP-hard, $\notin$ Poly-APX [6]
	Algorithm	Multi-LP	Spatial branch-and-bound [5, 6]	Multi-lex-MILP	–
$n \geq 4$ (fixed)	Complexity	P	NP-hard, $\notin$ Poly-APX [6]	NP-hard, $\notin$ Poly-APX	NP-hard, $\notin$ Poly-APX [6]
	Algorithm	Multi-LP	Spatial branch-and-bound [5, 6]	Multi-lex-MILP	–
$n \geq 4$ (free)	Complexity	NP-hard, $\notin$ Poly-APX	NP-hard, $\notin$ Poly-APX [6]	NP-hard, $\notin$ Poly-APX	NP-hard, $\notin$ Poly-APX [6]
	Algorithm	Multi-LP	Spatial branch-and-bound [5]	Multi-lex-MILP	–

lem is NP-hard with two or more followers, while, when the number of followers is three or more, the problem cannot be approximated in polynomial time to within any polynomial multiplicative factor unless P = NP (i.e., in formal terms, it is *not* in the class Poly-APX unless P = NP). To establish these two results, we introduce two reductions, one from Independent Set and the other one from 3-SAT.

After analyzing the complexity of the problem, we focus on its algorithmic aspects. First, we formulate the problem as a *bilevel programming problem*. We then show how to recast it as a single-level *quadratically constrained quadratic program* (QCQP),

Table 10.2 Summary of the results on the computation of Stackelberg equilibria in Stackelberg singleton congestion games with a single leader

Strong Stackelberg-Nash equilibrium

Leader's commitment			Pure	Mixed
Identical action spaces (symmetric games)	Monotonic costs	Complexity	P	P
		Algorithm	Greedy	Greedy
	Generic costs	Complexity	P	NP-hard, $\notin$ Poly-APX
		Algorithm	Dynamic programming	MILP
Different action spaces	Monotonic costs	Complexity	NP-hard, $\notin$ Poly-APX	NP-hard, $\notin$ Poly-APX
		Algorithm	MILP	MILP
	Generic costs	Complexity	NP-hard, $\notin$ Poly-APX	NP-hard, $\notin$ Poly-APX
		Algorithm	MILP	MILP

Weak Stackelberg-Nash equilibrium

Leader's commitment			Pure	Mixed
Identical action spaces (symmetric games)	Monotonic costs	Complexity	P	P
		Algorithm	Greedy	Greedy
	Generic costs	Complexity	P	NP-hard, $\notin$ Poly-APX
		Algorithm	Dynamic programming	Multi-lex-MILP
Different action spaces	Monotonic costs	Complexity	NP-hard, $\notin$ Poly-APX	NP-hard, $\notin$ Poly-APX
		Algorithm	Multi-lex-MILP	Multi-lex-MILP
	Generic costs	Complexity	NP-hard, $\notin$ Poly-APX	NP-hard, $\notin$ Poly-APX
		Algorithm	Multi-lex-MILP	Multi-lex-MILP

which we show to be impractical to solve due to admitting a supremum, but not a maximum. We then introduce a restriction based on a *mixed-integer linear program* (MILP) which, while forsaking optimality, always admits an optimal (restricted) solution. Next, we propose an exact algorithm to compute the value of the supremum of the problem based on an enumeration scheme which, at each iteration, solves a *lexicographic* MILP (lex-MILP) where the two objective functions are optimized in sequence. Subsequently, we embed the enumerative algorithm within a branch-and-bound scheme, obtaining an algorithm which is, in practice, much faster. We also extend the algorithms so that, for cases where the supremum is not a maximum, they return a strategy by which the leader can obtain a utility within an additive loss α with respect to the supremum, for any $\alpha > 0$. To conclude, we experimentally evaluate the scalability of our methods over a testbed of randomly generated instances.

A preliminary version of our results on normal-form Stackelberg games appeared in [17], while a complete version is [18].

10.3.2 Stackelberg Polymatrix Games

We identify two classes of Stackelberg polymatrix games that allow to characterize the complexity of computing Stackelberg-Nash equilibria (with followers restricted to pure strategies). The key property of these games is that, once fixed the number of players, computing a strong or weak equilibrium presents the same complexity, namely polynomial (again assuming that the followers play pure strategies). These games are of practical interest in security problems. Moreover, they are equivalent to *Bayesian* Stackelberg games with one leader and one follower, where the latter may be of different types. Our first class is equivalent to games with *interdependent types*, while the second one is equivalent to games with *independent types* (i.e., the leader's utility is independent of the follower's type). Thus, every result that holds for a game class also holds for its equivalent class.

We investigate whether the problem keeps being easy when the number of players is *not* fixed. We show that it is NP-hard to compute a weak Stackelberg-Nash equilibrium, and we provide an exact (exponential-time) algorithm (conversely, to compute a strong equilibrium, one can adapt the algorithm provided in [20] for Bayesian games, by means of our mapping). We also prove that, in all the instances where the weak Stackelberg-Nash equilibrium is a supremum but not a maximum, an α-approximation of the supremum can be found in polynomial time (also in the number of players) for any given additive loss $\alpha > 0$. As for approximation complexity, we show that the problem is Poly-APX-complete. This also shows that, in Bayesian Stackelberg games with uncertainty over the follower, computing a weak Stackelberg-Nash equilibrium is as hard as finding a strong one [33].

Next, we investigate whether, in general polymatrix games with followers restricted to play pure strategies, the problem admits polynomial-time approximation algorithms. We provide a negative answer, showing that in the strong case the problem is *not* in Poly-APX if the number of players is non-fixed, unless P $=$ NP. We also prove that the same inapproximability result holds for the weak case, even with a fixed number of players.

Our detailed results on Stackelberg polymatrix games appeared in [21] (see [22] for an extended version).

10.3.3 Stackelberg Congestion Games

We provide a comprehensive study of the computational complexity of finding Stackelberg-Nash equilibria in congestion games. These are games with a large number of players that compete for the use of some shared resources, where the cost of each resource is a function of the number of players using that resource, i.e., its *congestion*. Notice that, in such setting, assuming that the followers play a pure-strategy Nash equilibrium is without loss of generality, as congestion games always admit one [43].

First, we focus on games with *singleton* actions, i.e., where each player selects only one resource at a time. We draw a complete picture of the computational complexity of the problem of finding equilibria in Stackelberg singleton congestion games, with pure or mixed-strategy commitments, and considering the cases of finding either a strong equilibrium or a weak one. Interestingly, we identify two features which allow for thoroughly characterizing hard and easy game instances. The first one concerns the relationship among the action spaces of the players, with two possibilities: the one where the players are *symmetric* as they have identical action spaces and therefore they share the same set of resources, and the one where their action spaces may differ. The second feature is related to the shape of the players' cost functions. Two cases are possible: the one where these functions are *monotonically increasing* in the resource congestion and the one in which they may be not.

In particular, we show that, in games where the players' action spaces can be different, computing a (strong or weak) Stackelberg-Nash equilibrium is *not* in Poly-APX unless $P = NP$ even when the players' cost functions are monotonic, the leader has only one action available, and her costs are equal to the followers'. This result also holds if we restrict the leader to pure-strategy commitments, given that the leader has only one action available. For symmetric games where the players have identical action spaces, we show that the complexity of computing an equilibrium depends on the nature of the players' cost functions. For the case where the players' costs are generic (monotonic or not) functions of the resource congestion, we prove that the problem is *not* in Poly-APX unless $P = NP$. On the other hand, we show that, in symmetric games, the problem of computing a strong or weak Stackelberg-Nash equilibrium can be solved in polynomial time when the cost functions are monotonic by proposing an algorithm for it. We also consider the case where the leader is restricted to pure-strategy commitments, providing a polynomial-time algorithm for its solution which applies even to symmetric games with generic cost functions. This algorithm is based on a polynomial-time dynamic programming algorithm available in the literature for computing a socially optimal Nash equilibria in non-Stackelberg singleton congestion games with identical action spaces, which we improve and extend to solve our problem.

Then, we switch the attention to games beyond singleton ones. We show that having actions made of only one resource is necessary to have efficient (polynomial-time) algorithms. Indeed, we prove that finding a strong Stackelberg-Nash equilibrium is NP-hard and *not* in Poly-APX unless $P = NP$, even if players' actions contain only two resources, costs are monotonic, and players are symmetric. We also introduce and study singleton congestion games in which the players are partitioned into *classes*, with followers of the same class sharing the same set of actions. These are a generalization of singleton games with symmetric players, capturing the common case in which users can be split into (usually few) different classes, such as, e.g., users with different priorities. For these games, we provide a dynamic programming algorithm that computes a strong Stackelberg-Nash equilibrium in polynomial time, when the number of classes is fixed and the leader is restricted to play pure strategies. On the other hand, we prove that, if the leader is allowed to play mixed strategies, then the problem becomes NP-hard even with only four classes and monotonic costs.

Finally, for all the settings we study, we design MILP formulations for computing a strong Stackelberg-Nash equilibrium, and we experimentally evaluate them on a testbed containing both randomly generated game instances and worst-case instances based on our hardness reductions.

The results related to singleton games appeared in [36] and its extended version [15]. Instead, all the other results are provided by [34] (see [35] for an extended version).

10.4   Stackelberg Games with Multiple Leaders

We study games with multiple leaders, providing a new way to apply the Stackelberg paradigm to any finite (underlying) game. Our approach extends the idea of commitment to *correlated strategies* in settings involving multiple leaders and followers, generalizing the work of [19]. The crucial component of our framework is that a leader can decide whether to participate in the commitment or to defect from it by becoming a follower. This induces a preliminary *agreement stage* that takes place before the underlying game is played, where the leaders decide, in turn, whether to opt out from the commitment or not. We model this stage as a sequential game, whose size is factorial in the number of players. Our goal is to identify commitments guaranteeing some desirable properties on the agreement stage. The first one requires that the leaders do not have any incentive to become followers. It comes in two flavors, called *stability* and *perfect stability*, which are related to, respectively, Nash and subgame perfect equilibria of the sequential game of the agreement stage. The second property is also defined in two flavors, namely *efficiency* and *perfect efficiency*, both enforcing Pareto optimality with respect to the leaders' utility functions, though at different levels of the agreement stage.

We introduce three solution concepts, which we generally call *Stackelberg correlated equilibria*. They differ depending on the properties they call for. Specifically, (simple) Stackelberg correlated equilibria, Stackelberg correlated equilibria *with perfect agreement*, and Stackelberg correlated equilibria *with perfect agreement and perfect efficiency* require, respectively, stability and efficiency, perfect stability and efficiency, and both perfect stability and perfect efficiency.

First, we investigate the game theoretic properties of our solution concepts. We show that Stackelberg correlated equilibria with or without perfect agreement are guaranteed to exist in any game, while Stackelberg correlated equilibria with perfect agreement and perfect efficiency may not. Moreover, we compare the former with other solution concepts, both Stackelberg and non-Stackelberg ones.

Then, we switch the attention to the computational complexity perspective. We show that, provided a suitably defined *stability oracle* is solvable in polynomial time, a Stackelberg correlated equilibrium optimizing some linear function of leaders' utilities (such as the leaders' social welfare) can be computed in polynomial time, even in the number of players. The same holds for finding *a* Stackelberg correlated equilibrium with perfect agreement, while we prove that computing an optimal one is an intractable problem. Nevertheless, in the latter case, we provide an (exponential in the game size) upper bound on the necessary number of queries to the oracle.

In conclusion, we study which classes of games admit a polynomial-time stability oracle, focusing on *succinct games of polynomial type* [41]. The problem solved by our oracle is strictly connected with the *weighted deviation-adjusted social welfare problem* introduced by [28]. As a result, we get that our oracle is solvable in polynomial time in all game classes where the same holds for finding an optimal correlated equilibrium.

Our results on Stackelberg games with multiple leaders appeared in [13] (see [14] for an extended version).

10.5   Trembling-Hand Perfection in Stackelberg Games

We study Stackelberg games with a sequential structure, usually referred to as *extensive-form* Stackelberg games. In particular, we show that classical Stackelberg equilibria may prescribe the players to play sub-optimally off the equilibrium path, as it is the case for the Nash equilibrium. Thus, in order to amend these weaknesses, we propose a way to refine Stackelberg equilibria thorough trembling-hand perfection, which is based on the idea that each player might play each action with low-but-non-zero probabilities, usually called *trembles* [45].

We show that for every perturbation scheme (i.e., any possible way of introducing trembles), the set of limit points of Stackelberg equilibria for perturbed games with vanishing perturbations is always a nonempty subset of the Stackelberg equilibria of the non-perturbed game. This does not hold when focusing only on strong (or weak) equilibria: for a given game, the set of strong Stackelberg equilibria (or weak Stackelberg equilibria) in the non-perturbed game may be disjoint from the set of limit points of strong Stackelberg equilibria (or weak Stackelberg equilibria) in the perturbed game. We resort to the perturbation schemes used for *quasi-perfect* equilibria [48] and *extensive-form* perfect equilibria [45] to define their Stackelberg counterpart—and their strong and weak versions—as refinements of the Stackelberg equilibrium.

Next, we focus on quasi-perfection. We formally define the *quasi-perfect Stackelberg equilibrium* refinement game theoretically in the same axiomatic fashion as the quasi-perfect equilibrium was defined for non-Stackelberg games [48]. Thus, our definition is based on a set of properties of the players' strategies, and it cannot be directly used to search for a quasi-perfect Stackelberg equilibrium. Subsequently, we define a class of perturbation schemes for the sequence form such that any limit point of a sequence of Stackelberg equilibria in perturbed games with vanishing perturbation is a quasi-perfect Stackelberg equilibrium. This class of perturbation schemes strictly includes those used to find a quasi-perfect equilibrium by [40]. Then, we extend the algorithm by [16] to the case of quasi-perfect Stackelberg equilibrium computation. We derive the corresponding mathematical program for computing a *Stackelberg extensive-form correlated equilibrium* when a perturbation scheme is introduced and we discuss how the individual steps of the algorithm change. In particular, the implementation of our algorithm is much more involved, requiring the combination of branch-and-bound techniques with arbitrary-precision arithmetic to deal with small perturbations. This does not allow a direct application of off-the-shelf solvers. Finally, we experimentally evaluate the scalability of our algorithm.

In conclusion, we also study the computational complexity of finding Stackelberg equilibrium refinements, showing that the problem of deciding the existence of a Stackelberg equilibrium—refined or not—that gives the leader expected value at least v is NP-hard.

Our results appeared in [25] and [38] (see [37] for an extended version).

References

1. Amaldi E, Capone A, Coniglio S, Gianoli LG (2013) Network optimization problems subject to max-min fair flow allocation. IEEE Commun Lett 17(7):1463–1466
2. An B, Pita J, Shieh E, Tambe M, Kiekintveld C, Marecki J (2011) Guards and protect: next generation applications of security games. ACM SIGecom Exchanges 10(1):31–34
3. Avenhaus R, Okada A, Zamir S (1991) Inspector leadership with incomplete information. In: Game equilibrium models IV, pp 319–361. Springer, Berlin
4. Basilico N, Coniglio S, Gatti N (2017) Methods for finding leader–follower equilibria with multiple followers. arXiv:1707.02174
5. Basilico N, Coniglio S, Gatti N, Marchesi A (2017) Bilevel programming approaches to the computation of optimistic and pessimistic single-leader-multi-follower equilibria. In: SEA, vol 75, pp 1–14. Schloss Dagstuhl-Leibniz-Zentrum fur Informatik GmbH. Dagstuhl Publishing, Wadern
6. Basilico N, Coniglio S, Gatti N, Marchesi A (2019) Bilevel programming methods for computing single-leader-multi-follower equilibria in normal-form and polymatrix games. EURO J Comput Optim 8:1–29
7. von Stengel B, Zamir S (2010) Leadership games with convex strategy sets. Games Econ Behav 69(2):446–457
8. Bošanský B, Cermak J (2015) Sequence-form algorithm for computing stackelberg equilibria in extensive-form games. In: Proceedings of the twenty-ninth AAAI conference on artificial intelligence (AAAI 2015), pp 805–811 (2015)
9. Brown N, Sandholm T (2018) Superhuman ai for heads-up no-limit poker: libratus beats top professionals. Science 359(6374):418–424
10. Brown N, Sandholm T (2019) Superhuman ai for multiplayer poker. Science 365(6456):885–890
11. Murray C, Joseph Hoane A Jr, Hsu F-H (2002) Deep blue. Artif Intell 134(1–2):57–83
12. Caprara A, Carvalho M, Lodi A, Woeginger GJ (2016) Bilevel knapsack with interdiction constraints. INFORMS J Comput 28(2):319–333
13. Castiglioni M, Marchesi A, Gatti N (2019) Be a leader or become a follower: the strategy to commit to with multiple leaders. In: Proceedings of the twenty-eighth international joint conference on artificial intelligence, IJCAI 2019, Macao, China, August 10–16, 2019, pp 123–129
14. Castiglioni M, Marchesi A, Gatti N (2019) Be a leader or become a follower: the strategy to commit to with multiple leaders (extended version). CoRR, arxiv:abs/1905.13106

15. Castiglioni M, Marchesi A, Gatti N, Coniglio S (2019) Leadership in singleton congestion games: what is hard and what is easy. Artif Intell 277:103177
16. Cermak J, Bošanský B, Durkota K, Lisý V, Kiekintveld C (2016) Using correlated strategies for computing stackelberg equilibria in extensive-form games. In: Proceedings of the thirtieth AAAI conference on artificial intelligence (AAAI 2016)
17. Coniglio S, Gatti N, Marchesi A (2017) Pessimistic leader-follower equilibria with multiple followers. In: Proceedings of the twenty-sixth international joint conference on artificial intelligence, IJCAI-17, pp 171–177
18. Stefano C, Nicola G, Alberto M (2020) Computing a pessimistic stackelberg equilibrium with multiple followers: the mixed-pure case. Algorithmica 82:1189–1238
19. Conitzer V, Korzhyk D (2011) Commitment to correlated strategies. In: Proceedings of the twenty-fifth AAAI conference on artificial intelligence (AAAI 2011), pp 632–637
20. Conitzer V, Sandholm T (2006) Computing the optimal strategy to commit to. In: Proceedings of the 7th ACM conference on electronic commerce, pp 82–90
21. De Nittis G, Marchesi A, Gatti N (2018) Computing the strategy to commit to in polymatrix games. In: Proceedings of the thirty-second AAAI conference on artificial intelligence (AAAI-18), pp 989–996
22. De Nittis G, Marchesi A, Gatti N (2018) Computing the strategy to commit to in polymatrix games (extended version). CoRR, arxiv:abs/1807.11914
23. Farina G, Gatti N (2017) Adopting the cascade model in ad auctions: efficiency bounds and truthful algorithmic mechanisms. J Artif Intell Res 59:265–310
24. Farina G, Gatti N, Sandholm T (2018) Practical exact algorithm for trembling-hand equilibrium refinements in games. In: Advances in neural information processing systems, pp 5039–5049
25. Farina G, Marchesi A, Kroer C, Gatti N, Sandholm T (2018) Trembling-hand perfection in extensive-form games with commitment. In: Proceedings of the twenty-seventh international joint conference on artificial intelligence, IJCAI 2018, July 13–19, 2018, Stockholm, Sweden, pp 233–239
26. Gan J, Elkind E, Wooldridge M (2018) Stackelberg security games with multiple uncoordinated defenders. In: Proceedings of the 17th international conference on autonomous agents and multiagent systems, pp 703–711. International Foundation for Autonomous Agents and Multiagent Systems, Richland, SC
27. Gatti N, Lazaric A, Rocco M, Trovò F (2015) Truthful learning mechanisms for multi-slot sponsored search auctions with externalities. Artif Intell 227.93–139
28. Xin Jiang A, Leyton-Brown K (2011) A general framework for computing optimal correlated equilibria in compact games. In: International workshop on internet and network economics, pp 218–229. Springer, Berlin
29. Kiekintveld C, Jain M, Tsai J, Pita J, Ordóñez F, Tambe M (2009) Computing optimal randomized resource allocations for massive security games. In: AAMAS, pp 689–696
30. Labbé M, Marcotte P, Savard G (1998) A bilevel model of taxation and its application to optimal highway pricing. Manag Sci 44(12-part-1):1608–1622
31. Labbé M, Violin A (2016) Bilevel programming and price setting problems. Ann Operat Res 240(1):141–169
32. Letchford J, Conitzer V (2010) Computing optimal strategies to commit to in extensive-form games. In: Proceedings of the 11th ACM conference on electronic commerce, pp 83–92. ACM, New Year
33. Letchford J, Conitzer V, Munagala K (2009) Learning and approximating the optimal strategy to commit to. In: International symposium on algorithmic game theory, pp 250–262. Springer, Berlin
34. Marchesi A, Castiglioni M, Gatti N (2019) Leadership in congestion games: multiple user classes and non-singleton actions. In: Proceedings of the twenty-eighth international joint conference on artificial intelligence, IJCAI 2019, Macao, China, August 10–16, 2019, pp 485–491
35. Marchesi A, Castiglioni M, Gatti N (2019) Leadership in congestion games: multiple user classes and non-singleton actions (extended version). CoRR, arxiv:abs/1905.13108

36. Marchesi A, Coniglio S, Gatti N (2018) Leadership in singleton congestion games. In: Proceedings of the twenty-seventh international joint conference on artificial intelligence, IJCAI 2018, July 13–19, 2018, Stockholm, Sweden, pp 447–453
37. Marchesi A, Farina G, Kroer C, Gatti N, Sandholm T (2018) Quasi-perfect stackelberg equilibrium. CoRR, arxiv:abs/1811.03871
38. Marchesi A, Farina G, Kroer C, Gatti N, Sandholm T (2019) Quasi-perfect stackelberg equilibrium. In: The thirty-third AAAI conference on artificial intelligence, AAAI 2019, Honolulu, Hawaii, USA, January 27–February 1, 2019, pp 2117–2124
39. Matuschke J, McCormick ST, Oriolo G, Peis B, Skutella M (2017) Protection of flows under targeted attacks. Oper Res Lett 45(1):53–59
40. Miltersen PB, Sørensen TB (2010) Computing a quasi-perfect equilibrium of a two-player game. Econ Theory 42(1):175–192
41. Papadimitriou CH, Roughgarden T (2008) Computing correlated equilibria in multi-player games. J ACM (JACM) 55(3):14
42. Paruchuri P, Pearce JP, Marecki J, Tambe M, Ordonez F, Kraus S (2008) Playing games for security: an efficient exact algorithm for solving bayesian stackelberg games. In: Proceedings of the 7th international joint conference on Autonomous agents and multiagent systems, pp 895–902
43. Rosenthal RW (1973) A class of games possessing pure-strategy nash equilibria. Int J Game Theory 2(1):65–67
44. Sandholm WH (2002) Evolutionary implementation and congestion pricing. Rev Econ Stud 69(3):667–689
45. Selten R (1975) Reexamination of the perfectness concept for equilibrium points in extensive games. Int J Game Theory 4(1):25–55
46. Silver D, Huang A, Maddison CJ, Guez A, Sifre L, Van Den Driessche G, Schrittwieser J, Antonoglou I, Panneershelvam V, Lanctot M, et al (2016) Mastering the game of go with deep neural networks and tree search. Nature 529(7587):484
47. Milind T (2011) Security and game theory: algorithms, deployed systems, lessons learned. Cambridge University Press, Cambridge
48. Van Damme E (1984) A relation between perfect equilibria in extensive form games and proper equilibria in normal form games. Int J Game Theory 13(1):1–13
49. Van Damme E (1987) Stability and perfection of nash equilibria. Springer, Berlin, Heidelberg
50. Von Stackelberg H (1934) Marktform und gleichgewicht. Verlag von Julius Springer, Berlin

Open Access This chapter is licensed under the terms of the Creative Commons Attribution 4.0 International License (http://creativecommons.org/licenses/by/4.0/), which permits use, sharing, adaptation, distribution and reproduction in any medium or format, as long as you give appropriate credit to the original author(s) and the source, provide a link to the Creative Commons license and indicate if changes were made.

The images or other third party material in this chapter are included in the chapter's Creative Commons license, unless indicated otherwise in a credit line to the material. If material is not included in the chapter's Creative Commons license and your intended use is not permitted by statutory regulation or exceeds the permitted use, you will need to obtain permission directly from the copyright holder.

Chapter 11
Advancing Joint Design and Operation of Water Resources Systems Under Uncertainty

Federica Bertoni

11.1 Introduction

Hydropower has been employed as the first renewable energy source for electricity generation back in the 19th century and today it still plays a major, multidimensional role in the electricity sector worldwide for a variety of reasons. Firstly, it is a clean and renewable source of energy that generates local, affordable power fostering sustainable development, as promoted under the Sustainable Development Goals (SDGs) [1]. Secondly, it allows to reduce dependence upon imported fuels, associated to high risks of price volatility and supply uncertainty. Then, hydropower dams can offer multiple co-benefits, from storing water for drinking and irrigation, to being used for drought-preparedness, flood mitigation and recreation. In the end, hydropower is very competitive with other electricity sources from a costs point of view and provides a rapid-response when intermittent energy sources (e.g., solar) are off-line [2, 3].

Hydropower is currently responsible for about 16% of global electricity production, a percentage that is projected to substantially increase due to the doubling of the total installed hydropower capacity expected by 2050 [4]. Since developed countries already exploited more than 50% of their hydropower potential, most of the future hydropower expansion is predicted to occur in developing countries, which still present a vast untapped potential. Among others, Africa represents an extreme case with its almost 90% of undeveloped hydropower potential, with respect to a 25% global exploitation rate on average [2, 5]. This has motivated potential investments in the construction of approximately 3,700 new dams in the near future [6], a large share of which will be built in Africa, Asia and Latin America [4], leading to potential benefits in terms of increased energy supply but also negative impacts on the environment (e.g., losses of fish biodiversity, deforestation).

F. Bertoni (✉)
Politecnico di Milano, Piazza Leonardo da Vinci, 32, 20133 Milano, Italy
e-mail: federica.bertoni@polimi.it

© The Author(s) 2021
A. Geraci (ed.), *Special Topics in Information Technology*,
PoliMI SpringerBriefs, https://doi.org/10.1007/978-3-030-62476-7_11

In the future, changes in water availability and extreme events (e.g., droughts) due to climate change coupled with high rates of population growth will contribute to an increase in both migration rates within a single region, as well as energy and food demands, putting additional pressures on already stressed water resources. Globally, both existing and planned dams will thus have to face a vast array of future challenges, such as water scarcity, and growing resource conflicts in their demands (e.g., hydropower production vs irrigation water supply).

When planning new dams, integrated, strategic approaches must be therefore employed to find a balance between key economical, social and environmental objectives, while accounting for different water users and changes in external conditions that might strongly impact water resources systems in the future.

11.1.1 Research Challenges

The planning of large dams traditionally consists in basin-wide assessments of the potential economic outcomes of different designs via financial metrics (e.g., net present value) to evaluate their corresponding financial value [7, 8]. This approach combines costs and monetized downstream impacts of large water infrastructures into a single aggregate monetary value, disregarding potentially conflicting objectives and trade-offs among different water users within the basins of focus.

Secondly, over the last fifty years the interdependency between dam size and operations has been largely neglected by traditional engineering approaches relying on the widespread Rippl method [9], aimed at identifying a single optimal dam size based on a sequence of pre-defined releases and observed inflows [10, 11].

Third, the long design life of large dams critically exposes them to future uncertainties related to climatic and socio-economic changes. Yet, their planning is usually performed assuming stationarity in the long-term natural processes and without accounting for uncertainty in the external drivers. Since the assumption of a stationary climate is unlikely to be valid in the future [12], uncertainties in the main external drivers must be taken into account during dam planning in order to design robust infrastructures that are able to perform satisfactorily in the future with respect to multiple sources of uncertainty.

Building on the above mentioned research challenges, this contribution proposes a set of modelling and optimization tools converging in multiple, novel integrated frameworks for thoroughly capturing interdependencies between planning and operation in non-linear systems, also with respect to uncertainty in the main external drivers (e.g., hydro-climatology, human demands). The main focus is on water resources systems and specifically on coupling dam sizing and operation design. In particular, Sect. 11.2 presents a novel Reinforcement Learning (RL)-based approach to integrate dam sizing and operation design, while significantly containing computational costs with respect to alternative state-of-the-art methods. On the other hand, Sect. 11.3 shows a novel framework combining Multi-Objective Robust Decision Making and Evolutionary Multi-Objective Direct Policy Search into a novel approach to dam sizing, which internalizes the operation design problem and explicitly considers uncertainty in external drivers.

11.2 Reinforcement Learning for Designing Water Reservoirs

The method proposed relies on a novel algorithm, called Planning Fitted Q-Iteration (pFQI), which extends the batch-mode RL Fitted Q-Iteration (FQI) algorithm developed by [13] by enlarging the original FQI state space to include the discrete planning decision (i.e., dam size) as an additional state variable. The key idea behind pFQI originates from the Multi-Objective Fitted-Q Iteration (MOFQI) algorithm developed by [14]: the continuous approximation of the action-value function originally performed by FQI over the state-action space is now enlarged to the planning space by including dam sizes as new variables within the arguments of the action-value function. This enables pFQI to approximate the optimal operating policy associated to any dam size within a single learning process.

This new algorithm therefore overcomes the limitations and biases introduced by traditional sizing methods by directly addressing the strict interdependency between dam size and operation within an integrated framework through an operating policy parametric in the dam size. Secondly, it overcomes the high computational costs associated with state-of-the-art nested and integrated approaches by solving a single operation optimization via pFQI, as the resulting policy can be used to simulate the optimal operations of all the possible dam sizes associated to alternative trade-offs between least cost planning and operating objectives (e.g., downstream water supply). This characteristic contributes to significantly reducing the computational burden of pFQI.

The pFQI algorithm is tested on the multi-objective numerical case study, consisting of a synthetic water reservoir that must be sized and simultaneously operated to satisfy the water demand of downstream users.

11.2.1 pFQI Algorithm

The novel principle underlying the pFQI algorithm is to enlarge the traditional FQI state-action (x_t, u_t) space to include an additional time invariant state variable, namely the dam size $\theta \in \Theta$. This latter is described by the dummy deterministic state transition function $\theta_{t+1} = \theta_t = \theta$, where the dam size does not evolve in time, assuming a constant value throughout the entire evaluation horizon. The key idea behind pFQI originates from the Multi-Objective Fitted Q-Iteration (MOFQI) algorithm developed by [14], where linear combinations of preferences (weights) assigned to the objectives represent the dummy state variable to generate an entire Pareto front in a single optimization run. The resulting enlarged state-action space of pFQI can be therefore defined as $(\tilde{x}_t, u_t)$, where $\tilde{x}_t = [x_t, \theta]$, over which the optimal Q-function is continuously approximated. A new learning dataset $\mathcal{F}_\theta$ is thus produced, enlarging the original set of experience samples $\mathcal{F}$ as follows:

$$\mathcal{F}_\theta = \left\{ < \tilde{x}_t^i, u_t^i, \tilde{x}_{t+1}^i, g_{t+1}^i >, \quad i = 1, \ldots, N_\theta \right\} \tag{11.1}$$

where $N_\theta = N \cdot n_\theta$ is the number of tuples in the new pFQI dataset and n_θ is the number of sampled dam sizes θ. Since n_θ new tuples are produced for each four-tuple in $\mathcal{F}$, N_θ is larger than N. A tabular version of the Planning FQI algorithm proposed in this study is presented in Algorithm 1.

Likewise the traditional FQI algorithm, pFQI exploits the information in the sample dataset $\mathcal{F}_\theta$ to iteratively approximate the optimal action-value function $Q^*(\tilde{x}_t, u_t)$ over the enlarged state-action space $(\tilde{x}_t, u_t)$. Being the optimal operating rule, and thus operating policy, strictly related to the Q-function, the pFQI algorithm is also able to learn a continuous approximation of the optimal operating policy $\pi^*(\cdot)$ over the enlarged state space $\tilde{x}_t = [x_t, \theta]$. This policy therefore results parametric in the dam size $\theta \in \Theta$, and can be used to operate any dam size within this feasibility set.

Algorithm 1 Planning FQI Algorithm

Inputs: a learning set of tuples $\mathcal{F}_\theta$ and a regression
algorithm
Initialization:
Set $h = 0$
Set $\hat{Q}_0(\cdot) = 0$ over the whole enlarged state-action space $\tilde{X} \times \mathcal{U}$
Iterations: repeat until stopping conditions are met
- $h \leftarrow h + 1$
- build the training set
$TS_\theta = \{(IN^i, OUT^i), i = 1, \ldots, N_\theta\}$
where
$IN^i = (\tilde{x}_t^i, u_t^i)$
$OUT^i = g_{t+1}^i + \gamma \min_{u_{t+1}} \hat{Q}_{h-1}(\tilde{x}_{t+1}^i, u_{t+1}^i)$
- Run the regression algorithm on TS_θ to get $\hat{Q}_h(\tilde{x}_t, u_t)$
Output: derive the final operating policy $\hat{\pi}_h^*$ parametric in θ

11.2.2 Comparison with Traditional Least Cost Dam Design

In order to prove the importance of capturing interdependencies between dam size and operations, we compare the least cost planning solutions identified via traditional sizing methods and the optimal system configurations designed via pFQI.

The traditional sizing method identifies the least cost dam size under a pre-defined operating policy $\bar{\pi}$ while meeting a specific reliability rate. The pre-defined operating policy $\bar{\pi}$ adopted is the Standard Operating Policy (SOP), which assumes that the system operator is able to fully supply the downstream demand, unless constrained by water availability in the reservoir storage and current period inflow [15]. We test four different reliability rates $\bar{r}$, namely 95% (acceptable value for a dam aimed at supplying water for agriculture according to [16] and [17]), 90%, 85% and 80%, which are associated to four alternative least cost dam sizes operated under the same given operating policy $\bar{\pi}$. This latter consists of a target release equal to the downstream water demand w to be discharged at each time step.

Figure 11.1 shows the performance of four alternative least cost dam sizes optimized under four different reliability rates $\bar{r}$ (pink circles), along with the performance of two specific dam sizes θ_{opt} and θ_{sim} attained via nested approach (yellow triangles) The four least cost dam sizes increase proportionally with the associated reliability. For each of them, the operating objective (i.e., water supply deficit) is computed assuming a constant release equaling the downstream water demand unless constrained by the physical water availability, namely the pre-defined Stan-

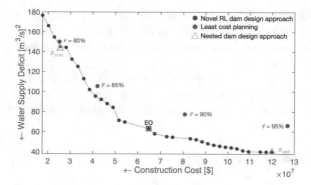

Fig. 11.1 2D-objective space where both the least cost planning solutions (pink circles) and the two nested solutions (yellow triangles) are compared against the system configurations identified via novel RL approach relying on the pFQI algorithm (blue circles). Arrows indicate the direction of preference in the objectives

dard Operating Policy $\bar{\pi}$. The four least cost system configurations are compared against the optimal dam sizes and associated operating policy identified via novel RL approach (blue circles). In this first experiment, the size of the learning dataset $\mathcal{F}_\theta$ of the pFQI algorithm employed during the novel RL operation optimization is $n_\theta = 11$. As can be observed, the RL solutions dominate the four least cost system configurations in terms of system performance. In particular, the dam size associated to the highest reliability rate $\bar{r} = 95\%$, which has a cost equal to \$ 127 million (106 Mm^3), is weakly dominated by the Equivalently Operated (EO) solution identified via novel RL approach ($J^c = \$ 65$ million and size equal to 54 Mm^3).

A 50% smaller (thus less costly) infrastructure could be built while achieving the same water supply deficit, meaning that the pFQI algorithm is able to fully capture the interdependency between dam size and operation within an integrated framework, yielding system configurations that strongly outperform the performance achieved under traditional sizing methods. This is particularly true for reliability rates that are equal to or higher than 85%, where the advantages of jointly optimizing dam size and operation can be seen in terms of a significant deviation between the least cost planning and the RL solutions. For lower reliability rates, the corresponding water supply deficit is so high that almost any operating policy is able to attain it. The potential benefits of optimizing the operation thus become almost negligible and the least cost system configuration associated to an 80% reliability approaches the RL solutions. By coupling smaller dam sizes to more efficient operating schemes, and accounting for the impacts of short-term operating policies on the long-term system design, the novel RL dam design approach is therefore able to identify less costly, more efficient system designs.

11.3 A Novel Robust Assessment Framework

In this section, we propose a robust dam design framework that is: (i) multi-objective; (ii) capable of joint planning and management, capturing the interdependencies between dam size and the associated trade-offs across candidate operating policies; (iii) integrating state-of-the-art stochastic optimization, yielding design system configurations that are less vulnerable to intrinsic, stationary hydro-climatic variability; and (iv) directly accounting for robustness to long term deep uncertainties, clarifying how alternative system configurations perform with respect to uncertain drivers (e.g., inflows and demands).

We carefully evaluate the potential of this framework through an ex-post analysis of the Kariba dam in the Zambezi river basin, which provides a rich and challenging opportunity to demonstrate the limitations of prior standard sizing approaches.

11.3.1 Methodological Approach

Our proposed integrated design and operation framework therefore captures (i) key trade-offs between water users, (ii) how dependencies in dam sizing and operations influence these trade-offs, (iii) how intrinsic, stationary hydro-climatic variability affects dam design, and (iv) the ultimate robustness of the water infrastructures designed. Drawing on the methodological taxonomy for robustness frameworks suggested by [18], we present in the following the four main elements of our proposed framework.

The first element is the generation of decision alternatives, namely alternative dam sizes with associated candidate operating policies, via multi-objective optimization under both historical and well-characterized, stationary streamflow uncertainty. It is a useful insight to discover how planning and management trade-offs evolve when moving from historical observation record to a better statistical representation of the internal variability of extremes. Put more simply, the available historical observation record by itself is a poor estimator of rare extremes that are potentially highly consequential to dam sizing and operations.

The second element is strongly linked to the search in the first element by how alternative states of the world (SOWs) are sampled and exploited in the overall analysis. Sampling strategies can be classified into three different groups: (i) historical records, representing a single SOW composed of the available observed time-series; (ii) stationary synthetic records, where each SOW is obtained by sampling well-characterized, stationary model for uncertain hydro-climatic factors; and (iii) deeply uncertain scenarios, where each SOW is generated by globally sampling a suite of deeply uncertain drivers. As discussed above, stationary synthetic records better capture system's stochastic hydrology, where the historical autocorrelation of recorded streamflows is preserved while better accounting for their internal variability. As for sampling deep uncertainties, the broader suite of SOWs used to stress test systems are drawn from non-stationary future scenarios of both hydro-climatic (inflows) and socio-economic (irrigation demand) factors. The SOWs belonging to the first two groups are employed during the optimal alternatives generation phase, whereas the latter are used in the a-posteriori robustness assessment of the candidate alternatives identified.

Once the candidate alternatives for design and operation are re-evaluated over the deeply uncertain SOWs, their robustness is assessed in terms of global domain criterion satisficing metric [19, 20], namely the percentage of SOWs satisfying pre-defined performance requirements (third element). As an innovation in this study's assessment of robustness, instead of pre-specifying acceptable performance thresholds, we map the satisficing robustness measure into the objective space, forming a multidimensional satisficing surface.

The last element of our methodology is a sensitivity analysis conducted to identify which deeply uncertain factors are the most responsible for failing specific performance criteria (i.e., factor mapping or scenario discovery).

11.3.2 Assessing Robustness of Alternatives for Changing Demands and Hydrology

Figure 11.2 provides a broader evaluation of the hydropower maximizing (LH, MH, and SH) and compromise alternatives (LC, MC, and SC) for large L, medium M and small S dam sizes with respect to deeply uncertain changes in demands and hydroclimatology. These solutions were re-evaluated over our deeply uncertain scenarios, composed of 143 sampled combinations of changes in mean streamflow and irrigation demands. Here, the robustness of the different system configurations is assessed in terms of a satisficing surface, which maps the satisficing robustness measure across the conflicting objectives space for any combination of either hydropower (Fig. 11.2a) or multivariate (Fig. 11.2b) thresholds for both hydropower and irrigation deficit. In the case of the multivariate thresholds, the color of each point forming the satisficing surface is given by the exact value of the satisficing robustness measure, namely the percentage of SOWs satisfying that specific performance requirement on multiple objectives simultaneously.

Figure 11.2a distinguishes the hydropower focused robustness of the baseline forensic solution (i.e., historical operating policy associated to the existing Kariba dam size) relative to the large (LH), medium (MH), and small (SH) alternatives attained using the stochastic joint optimization. It is very clear that the existing Kariba

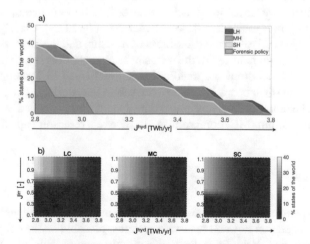

Fig. 11.2 Panel a: Cumulative distribution functions of the baseline forensic solution and the policy maximizing hydropower production H associated to three optimal dams sizes (S: Small, M: Medium, L: Large) across 143 deeply uncertain states of the world. Panel b: Mapping of the robustness in terms of satisficing metric of the compromise policy C in the 2D-management objective space. The color is given by the percentage of deeply uncertain states of the world that satisfy a specific multivariate threshold on both hydropower J^{hyd} and irrigation J^{irr} objectives (red = low percentage; green = high percentage). In both panels, arrows indicate the direction of preference in the objectives

design and operation does not in a general sense maximize hydropower. It has an extremely narrow hydropower production range (2.8–3.0 TWh/yr) where it meets performance goals for as few as 20% of sampled SOWs. Alternatively, Fig. 11.2a shows that the robustness of the large LH and medium MH dam sizes are similar to one another, as their cumulative distribution functions mostly overlap. This is due to the fact that the hydropower maximizing operating policies associated to these two dam alternatives behave the same, minimizing spillages and keeping releases constant regardless of the water level in the reservoir. Dam size starts playing an important role in robustness for small hydropower maximizing alternative (SH). Although this alternative spills large volumes of water due to its size limitations, it is striking that it fully outperforms the existing Kariba system being operated with the idealized bang-bang forensic solution, which aims at tracking a prescribed storage trajectory. Increasingly severe water scarcity captured in the deeply uncertain SOWs with respect to historical conditions causes the baseline forensic solution to fail as Kariba's operations are not able to drive the reservoir storage back to its prescribed storage trajectory. Note that the four cumulative distribution functions step-behavior is reflective of the discretized sampling of the deeply uncertain factor space across hydropower production levels (Fig. 11.2a).

The compromise alternatives (LC, MC, and SC) in Fig. 11.2b are evaluated for sustaining their performance in both the hydropower and irrigation deficit objectives for the more challenging deeply uncertain SOWs. It is clear in Fig. 11.2b that maintaining high levels of performance in both objectives is very difficult and, as a consequence, 0% of the sampled deeply uncertain SOWs meet the suite of performance goals for a large swath of mapped satisficing surface, regardless of dam size. However, there are some interesting differences across the dam sizes. Perhaps most surprisingly, the large compromise (LC) solution's robustness is not clearly superior to the medium size compromise (MC). The medium compromise (MC) alternative generalizes over lower irrigation deficits while still maintaining competitive performance in hydropower robustness (also noticeable in Fig. 11.2a). Clearly, volumetric size is not the sole control on the robustness of a design and the importance of operational policies are pronounced in the robustness results of Fig. 11.2b. Based on the solutions selected and discussed, medium (MC) and small (SC) compromise dam sizes are exploiting their operations to be more robust than the large (LC) compromise alternative. The small (SC) compromise dam is still disadvantaged volumetrically relative to the medium (MC) compromise alternative's reservoir capacity.

11.4   Conclusions

We presented two novel contributions whose main goal is to advance the current planning and operation of water reservoir systems, focusing on the coupling of dam sizing and operation design in order to thoroughly capture their interdependencies also with respect to uncertainty in the main external drivers.

Results show that capturing the interdependencies between dam size and operations has proved to be essential to effectively design smaller yet more efficient water reservoirs that strongly outperform the performance achieved under traditional engineering sizing methods, which instead neglect the optimal operation design phase. When employing our novel Reinforcement Learning (RL)-based approach on a numerical case study, where a synthetic reservoir must be sized and operated to meet downstream users water demand while minimizing construction costs, a 50% smaller (thus less costly) dam could be built without degrading the system performance achieved under the least cost infrastructure identified via traditional sizing method (i.e., Behavior Analysis). When instead we use our robust dam design framework to perform an ex-post analysis of the existing Kariba dam, which has been sized via traditional design methods assuming a pre-defined operating rule (i.e., target storage to be tracked), we are able to design a 32% less costly Kariba dam with respect to the existing one, without degrading system performance.

Secondly, a careful consideration of the broader array of future uncertainties within the planning phase is key for designing robust infrastructures, which will likely face a wide range of future challenges such as reduced water availability and rising frequency of extreme events (e.g., floods, droughts) related to climate change, together with increases in both energy and food demands due to population growth. By including stationary, hydro-climatic uncertainty within the Kariba dam design phase, we are able to identify well operated but reduced volume alternative reservoirs that can fully dominate larger designs in terms of their attained robustness. In particular, we clearly highlight that Kariba, even if optimally implementing its pre-defined operating rule, is critically vulnerable to stationary hydro-climatic variability, and that it produces less hydropower than a well sized run-of-the-river hydropower plant.

Future research should mainly focus on (i) including a broader array of external uncertainties in the coupled dam sizing and operation design problem in order to fully understand which uncertainties drive the robustness of water reservoir systems, by including them within the optimization process, and (ii) testing the contributed methodological approaches on complex transboundary, multi-reservoir systems, in order to explore potential interactions among several dams planned for the near future and that must be operated to satisfy different management objectives (e.g., environmental flow requirements).

References

1. United Nations (2019) Sustainable development goals knowledge platform. https://sustainabledevelopment.un.org
2. Cole MA, Elliott RJ, Strobl E (2014) Climate change, hydro-dependency, and the African dam boom. World Dev 60:84–98
3. International Finance Corporation (IFC) (2015) Hydroelectric power: a guide for developers and investors. The World Bank, Washington, DC
4. IEA (2012) Technology roadmap: hydropower. Organization for Economic Cooperation and Development, International Energy Agency, Paris
5. Berga L (2016) The role of hydropower in climate change mitigation and adaptation: a review. Engineering 2(3):313–318

6. Zarfl C, Lumsdon AE, Berlekamp J, Tydecks L, Tockner K (2015) A global boom in hydropower dam construction. Aquat Sci 77(1):161–170
7. Jeuland M (2009) Planning water resources development in an uncertain climate future: a hydro-economic simulation framework applied to the case of the Blue Nile. Ph.D. thesis, The University of North Carolina at Chapel Hill
8. Ray P, Kirshen P, Watkins D Jr (2011) Staged climate change adaptation planning for water supply in Amman, Jordan. J Water Resour Plan Manag 138(5):403–411
9. Rippl W (1883) The capacity of storage reservoirs for water supply. Van Nostrand's Eng Mag (1879–1886) 29(175):67
10. U.S. Army Corps of Engineers (1975) Hydrologic engineering methods for water resources development: reservoir yield, vol 8. Hydrologic Engineering Centre, Davis, CA
11. U.S. Army Corps of Engineers (1977) Hydrologic engineering methods for water resources development: reservoir system analysis for conservation, vol 9. Hydrologic Engineering Centre, Davis, CA
12. Milly PC, Betancourt J, Falkenmark M, Hirsch RM, Kundzewicz ZW, Lettenmaier DP, Stouffer RJ (2008) Stationarity is dead: whither water management? Science 319(5863):573–574
13. Ernst D, Geurts P, Wehenkel L (2005) Tree-based batch mode reinforcement learning. J Mach Learn Res 6(Apr): 503–556
14. Castelletti A, Pianosi F, Restelli M (2013) A multiobjective reinforcement learning approach to water resources systems operation: Pareto frontier approximation in a single run. Water Resour Res 49(6):3476–3486
15. McMahon TA, Adeloye AJ (2005) Water resources yield. Water Resources Publication
16. Nassopoulos H, Dumas P, Hallegatte S (2012) Adaptation to an uncertain climate change: cost benefit analysis and robust decision making for dam dimensioning. Clim Change 114(3–4):497–508
17. McMahon TA, Vogel RM, Pegram GG, Peel MC, Etkin D (2007) Global streamflows-part 2: reservoir storage-yield performance. J Hydrol 347(3–4):260–271
18. Herman JD, Reed PM, Zeff HB, Characklis GW (2015) How should robustness be defined for water systems planning under change? J Water Resour Plan Manag 141(10):04015012
19. Starr MK (1963) Product design and decision theory. Prentice-Hall, Upper Saddle River
20. Schneller G, Sphicas G (1983) Decision making under uncertainty: starr's domain criterion. Theor Decis 15(4):321–336

Open Access This chapter is licensed under the terms of the Creative Commons Attribution 4.0 International License (http://creativecommons.org/licenses/by/4.0/), which permits use, sharing, adaptation, distribution and reproduction in any medium or format, as long as you give appropriate credit to the original author(s) and the source, provide a link to the Creative Commons license and indicate if changes were made.

The images or other third party material in this chapter are included in the chapter's Creative Commons license, unless indicated otherwise in a credit line to the material. If material is not included in the chapter's Creative Commons license and your intended use is not permitted by statutory regulation or exceeds the permitted use, you will need to obtain permission directly from the copyright holder.

Chapter 12
Optimization-Based Control of Microgrids for Ancillary Services Provision and Islanded Operation

Alessio La Bella

12.1 Introduction

Climate change is today a well-known issue. The average world temperature is continuously rising, due to the high concentration of greenhouse gasses, and the negative effects on the environment are already more than evident. Among the involved causes, electricity production is one of the largest responsible, accounting for around one third of the global greenhouse gas emissions every year. Therefore, the actual centralized generation system, where most of the power is provided by fossil-based power plants, is no more a sustainable paradigm and this is today required to shift to a more distributed architecture, relying on a conspicuous spread of Renewable Energy Sources (RESs), as confirmed by the European Union in the Directive 2018/2001. The energy transition is not a straightforward path and many technical challenges must be tackled. A main issue is that renewable generation is intermittent and non-deterministic since it often depends on weather conditions, e.g. for wind-turbine and photovoltaic generators. This implies that RESs, by themselves, cannot guarantee the continuous balance between power generation and demand, which is a fundamental requirement not only to guarantee the secure power supply but also for the frequency' and voltages' stability [1]. A solution to overcome this issue is to combine the RESs' diffusion with a parallel integration of other distributed dispatchable resources, such as battery systems and smart modulable loads, which can be properly controlled to adapt their power patterns ensuring the power balance in the electrical system, i.e. providing the so-called *ancillary services* [2]. Nevertheless, this decentralized paradigm involves increasing coordination issues with respect to the traditional system, requiring the efficient management of the widespread distributed resources based on the real-time needs of the electrical system.

A. La Bella (✉)
Politecnico di Milano, Piazza Leonardo da Vinci 32, 20133 Milan, Italy
e-mail: alessio.labella@polimi.it

© The Author(s) 2021
A. Geraci (ed.), *Special Topics in Information Technology*,
PoliMI SpringerBriefs, https://doi.org/10.1007/978-3-030-62476-7_12

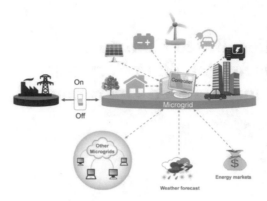

Fig. 12.1 Schematic of a generic MG [https://building-microgrid.lbl.gov]

In this context, *MicroGrids* (MGs) have been devised as a great solution to enhance the flexibility and controllability of the future electrical system. A MG is small-scale grid, usually large as a city district, incorporating RESs, batteries, micro-generators, and smart loads, which are then regulated by proper control architectures. The deployment of MGs would bring several benefits to the electrical system, as they allow a more effective management of RESs, thanks to the co-location with other dispatchable devices that can balance their variability [3]. Another significant advantage derives by the flexible nature of MGs, as they can be operated either grid-connected or in islanded mode. A schematic of a MG is shown in Fig. 12.1.

Nevertheless, MGs are still not recognized as a competitive alternative to fossil-based power plants. In fact, the electrical regulations on ancillary services provision are characterized by too high power requirements for single microgrids, as they have been traditionally requested to big power plants [4]. On the other hand, the MG islanded mode still involves several technical challenges due to the absence of the main utility's support. In fact, local resources must be controlled both to efficiently ensure the MG internal power balance, despite the presence of uncertain RESs, and to internally stabilize the MG frequency and voltages [5].

This doctoral thesis therefore have aimed to design several control architectures and algorithms to properly and efficiently operate MGs in grid-connected and islanded mode, overcoming the mentioned issues and so fostering the MGs' integration to facilitate the energy transition [6]. This summary brief is structured as follows. Section 12.2 provides an overview of different optimization-based control strategies for MGs to become valuable providers of ancillary services. Then, the designed control systems for islanded MGs are described in Sect. 12.3. Section 12.4 concludes this brief with some final remarks, describing possible future research directions.

12.2 Microgrids Aggregators Providing Ancillary Services

As mentioned in Sect. 12.1, MGs are usually characterized by a limited power capability, having to satisfy also their internal loads, and so they are not significant players for the ancillary service provision. To solve this issue, this doctoral thesis proposes several control architectures to jointly coordinate an aggregator of multiple interconnected MGs, properly regulated by an Aggregator Supervisor (AGS), so that they can act as a unique significant entity for the system operators, potentially capable of providing ancillary services [7]. Aggregation is in fact an effective solution to foster the participation of small-scale distributed resources to the ancillary services provision, as also recognized by the European Union in the Regulation 2017/2195 [8]. To properly accomplish this task, aggregators must perform the following operations:

- During the offline operation, e.g. the day-ahead, the AGS must define the optimal overall power exchange of the MGs Aggregator (MG-AG) with the main utility, considering the energy prices and the operation costs of the MGs' units. This phase must also consider the allocation of a proper amount of power reserve in the MG-AG, respecting the minimum requirements imposed by the regulations.
- During the online operation, e.g. in the intra-day, the AGS must be able to promptly reschedule MGs operations to compensate power imbalances in the electrical system, varying the pre-scheduled MG-AG output power as requested by the system operators.
- During the real-time operation, the AGS must ensure that the agreed MG-AG power profile is maintained despite the presence of non-deterministic RESs and loads inside the MG-AG, therefore all the internal unexpected imbalances must be promptly compensated.

These three phases will be the core of the control algorithms described in the next three paragraphs, respectively. It should be underlined that controlling a MG-AG is a large-scale and computationally-intensive problem and moreover, it is necessary to preserve MGs' internal information and the local control of their units, being them private facilities. Because of this, the proposed control algorithms are not defined according to pure centralized approaches, but novel distributed and hierarchical schemes are proposed with enhanced optimality and scalability properties, so that their performances are independent on the size of the aggregation.

12.2.1   Offline Economic Dispatch and Power Reserve Procurement

This offline phase aims to define the MG-AG optimal output power considering the energy prices and the internal MGs costs, as well as to procure a required minimum amount of power reserve. Additionally, MGs operations must be coordinated also to avoid over-voltage and congestion issues in the electrical network, which can be easily occur in presence of distributed generation.

To accomplish these tasks, firstly the optimal MG-AG power profiles are scheduled, considering economic objectives and the ancillary service provision. Then, in a second step, the electrical feasibility of the scheduled profiles is checked and modifications are implemented if necessary. For the first task, it should be considered that each MG wants to optimize the use of its own units, considering local cost functions and constraints. However, MGs optimization problems are all linked by coupling constraints related both to the MG-AG output power to sell/buy to/from the main utility and to the MG-AG power reserve that must be allocated. This implies that pure decentralized approaches can not be implemented as an interaction between the different MGs is needed. Because of this, this first step is addressed through the definition of a distributed optimization algorithm based on the *dual decomposition theory*, precisely the *Alternating Direction Method of Multipliers* (ADMM) [9]. Avoiding all the mathematical details of this approach, this technique allows to remove the constraints coupling the MGs optimization problems, expressing their violations as properly weighted costs through the so-called *Lagrangian relaxation* [10]. This makes the overall optimization problem separable, and so it can be distributed among the MGs and the AGS. The separable optimization problems are then solved through an iterative negotiation between the AGS and the MGs, which at convergence achieves the optimal solution and the feasibility of the coupling constraints. Precisely, the following sequential operations are iteratively executed:

- The AGS sends to the MGs properly defined internal prices, denoted as *dual variables*, for the MGs power output and allocated power reserve.
- MGs solve in parallel their optimization problems considering the AGS internal prices and their own units' costs. Then, they communicate to the AGS their optimal output power and allocated reserve.
- The AGS gathers this information and solves an internal optimization problem. Based on its optimal solution, it properly updates the *dual variables* trying to drive MGs towards the optimal and feasible solution.

The main advantage of the proposed approach is that MGs do not have to communicate their internal information to the AGS, as they directly optimize their internal units, and moreover MGs operate in parallel, implying that the overall computational time does not rise with their number. If some mild assumptions are respected (e.g. convexity of the optimization problems), the described algorithm converges to the same optimal objective of the centralized system, i.e. the case where the AGS has full knowledge and control of the MGs internal units. Having defined the optimal power profiles of the MG-AG, an additional optimization procedure is then performed to check power flow feasibility, considering the limitations on nodal voltages and line currents. This task is not easily solvable through a distributed approach since the corresponding optimization problem is inevitably non-convex, due to the power flow equations, and therefore convergence and feasibility issues arise [11]. Therefore, MGs are properly modeled as *equivalent generators*, which allows to address the electrical feasibility in a pure centralized way, without however requiring MGs to communicate their internal information and involving computational issues.

Figure 12.2 shows some results of the described procedure, considering as benchmark an aggregation of 4 MGs connected to the IEEE 37 bus system [12]. In particular, from Fig. 12.2a the optimal output power computed by the distributed algorithm is compared to the optimal solution of the centralized system, where it is evident that the same solution is achieved. Figure 12.2b shows instead the difference of the

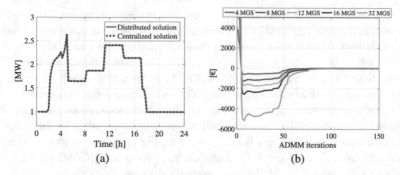

Fig. 12.2 **a** MG-AG optimal power output; **b** Optimality gap between centralized and distributed solution over the iterations of the proposed algorithm

optimal cost function between the centralized and the distributed solution over the number of iterations, considering also a varying number of MGs inside the MG-AG. As notable, the optimality gap always converges to zero in around 70 iterations, independently on the number of MGs, given the scalability properties of the proposed approach.

12.2.2   Online External Provision of Ancillary Services

During the online operation, the AGS has to reschedule the MG-AG operations as requested for the external provision of ancillary services. Precisely, the system operators can request the AGS to increase, or decrease, the MG-AG output power over a specific time period, defined as *request period*. Moreover, the AGS can also periodically offer additional power reserve to the system operators, if available.

The AGS has to respect additional constraints while performing these tasks. In fact, the MG-AG output power profile must be varied just during the request period, and not in the time instants subsequent it, avoiding the so-called *rebound effect* [13]. Moreover, neither the pre-agreed power reserve must be affected in the time instants subsequent to the request period, since ancillary services may be later requested.

It should be noted that MGs can decide to interrupt, shift or modulate some controllable loads to satisfy the system operators' requests. This results in the introduction of mixed-integer models, implying that standard distributed approaches may lead to sub-optimality and feasibility issues [14]. Therefore a novel hierarchical technique is proposed to perform this task, based on the definition of the *flexibility function* concept. These functions express the maximum and minimum power variation that each MG can provide over the request period, as well as the additional cost that each MG afford for any requested power variation. These become effective tools for the AGS to quickly, and optimally, reschedule the MG-AG operations as requested, and to offer additional power reserve if available. Moreover, also in this case MGs do not have to communicate their internal models and characteristics, but just this flexibility information. It is not simple to analytically characterize the MG flexibility functions, as they are inherently non-linear and non-convex due to the mixed-integer MGs modelling. However, it is shown in the doctoral thesis how each MG can easily compute a convex approximation of its flexibility function. This allows the AGS to be extremely efficient in computing the optimal solution, as it solves a convexification of the MG-AG rescheduling problem. The test results of this approach revealed to be particularly promising. Despite the introduced approximations, the optimality gap between the centralized solution and the solution computed by the proposed approach reached at maximum 0.1%. Moreover, the proposed technique revealed to be much faster than the centralized case in computing the MG-AG rescheduling, reaching a reduction of the 80% for the overall computational time.

12.2.3 Real-Time Self-balancing of Internal Power Uncertainties

The last operation concerns the tracking of the agreed MG-AG power profile, which is a critical task due to the presence of several non-controllable and non-deterministic RESs and loads inside the MG-AG, as their output power often deviate with respect to the forecasts. To overcome this issue, it is proposed to coordinate MGs to exploit the remaining power reserve, i.e. the one not externally requested by the system operators, compensating the internal power variability of the MG-AG. To perform this task, a scalable and prompt control architecture is required, as power imbalances must be quickly balanced even in large-scale electrical networks. This implies that neither centralized nor pure distributed approaches are advisable, as the former are not scalable, and the latter are not prompt but usually involve iterative procedures. Therefore, a novel control approach is here designed, which schematic is depicted in Fig. 12.3.

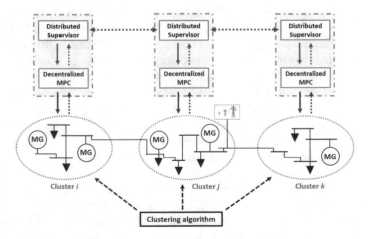

Fig. 12.3 Schematic of the proposed control architecture for real-time power balancing

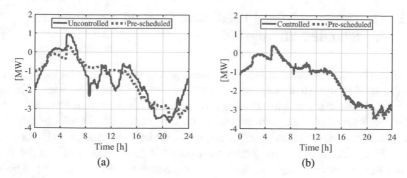

Fig. 12.4 MG-AG power exchange in the uncontrolled (**a**) and controlled case (**b**)

The first step of the approach is based on a properly designed *clustering procedure*, with the objective of creating network clusters that are as self-sufficient as possible in compensating unexpected internal power imbalances. The proposed clustering algorithm takes into account the remaining power reserves in MGs and the worst-case deviations of loads and RESs. Then, a two-layer control architecture is implemented. The lower layer is constituted by a decentralized framework of Model Predictive Control (MPC) regulators, each one managing a network cluster by coordinating the MGs to balance the cluster' power variability. However, if large power deviations occur in a cluster and the remaining power reserve is not sufficient to compensate it, the decentralized MPC regulator can issue a power request to the upper layer: the Distributed Supervisor. This is a fully distributed control scheme which defines the optimal power exchanges among the different clusters so that the overall network remains always balanced. This scheme is based on the *Distributed Consensus ADMM* algorithm, which allows the direct interaction between the different supervising agents [15].

The proposed architecture has been tested on several IEEE benchmarks, such as the 37, 118 and the 123 bus-systems. Figure 12.4 reports some results for the IEEE 37 bus-system, comparing the MG-AG power exchange with the main utility if the proposed control architecture is applied and if it is not. It is evident that the variability of loads and RESs seriously affects the MG-AG output power, as it significantly deviates from the pre-scheduled profile if imbalances are not promptly restored. On the other hand, the proposed scheme allows almost a perfect tracking of the pre-scheduled power profile, through the efficient exploitation of local MGs. It is worth noticing that the upper supervising layer, although distributed, is particularly fast in computing the optimal solution as most of the model complexity is addressed by the local MPC regulators. It is also worth noticing that the two control layers are decoupled, meaning that the MPC regulators continuously and autonomously operate to compensate all unexpected power deviations, while the supervisor is activated to optimally redefine the power exchanges among clusters just when necessary.

12.3 Hierarchical Model Predictive Control Architectures for Islanded Microgrids

A MG can be also operated in islanded mode, just relying on its own local sources. This is a valid solution to ensure power supply in critical facilities in case of blackout events (e.g. hospitals), and to electrify rural areas in a sustainable way, as in some developing countries where a consistent electric infrastructure is still absent. However, as mentioned, the islanded mode requires efficient and prompt architectures to control local resources, ensuring both the optimal MG management and the stabilization of the internal frequency and voltages. Since these tasks involve different time scales and system modelling, multi-layer control architectures are proposed considering MGs with alternating current networks (AC-MGs) and also MGs with direct current networks (DC-MGs), which have recently raised a significant interest for their high efficiency [16]. Among the proposed solutions, the three-layer control architecture for DC-MGs, designed during a visiting period at the Automatic Control Laboratory of EPFL, is here described. In case the reader is interested to the designed control architectures for islanded AC-MGs, please refers to [6, 17, 18].

The schematic of the proposed control architecture for islanded DC-MGs is depicted in Fig. 12.5. At the upper layer, a hybrid-MPC system is implemented, executed with a sampling time of 15 min, able to consider units' constraints, weather predictions, mixed-integer models and several economic aspects. The objective of this control layer is to ensure the optimal MG power balance, defining the generators' power references and switching on, or off, some MGs units when convenient. Then, low-level plug-and-play fast voltage controllers are designed, acting at the converter interfaces of the generation units, having the particular characteristic of guaranteeing the voltage stability also if generation units are switched on, or off. To properly

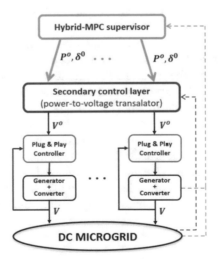

Fig. 12.5 Three-layer control architecture for an islanded DC-MG

interface these two layers, an innovative secondary optimization-based system is also designed, properly translating the power references provided by the hybrid MPC to the voltage references needed by the plug-and-play controllers. The theoretical properties of this interfacing layer have been analysed, due to its nonlinear and non-convex nature, and it has been proved that, under realistic and easy-verifiable assumptions, a solution for the power-to-voltage translation always exists and, moreover, this is unique [19]. These properties revealed to be particularly significant for ensuring the proper and secure operation of the whole architecture. The hierarchical control system has been tested on real DC-MG benchmark, showing its significant performances in ensuring the optimal and stable operation in many different conditions.

12.4 Conclusions

This chapter aimed to give an overview of the different control architectures designed for microgrids, considered the key-solution for enhancing the spread of renewable energy sources in the electrical system [6]. In particular, the control strategies presented in Sect. 12.2 showed to be really effective solutions to make microgrids fundamental players in the electrical system, as they can cooperate as part of a unique aggregator to efficiently provide ancillary services. Moreover, the hierarchical control architectures designed for islanded microgrids, briefly described in Sect. 12.3, allow the optimal energy management of the local sources and to stabilize the internal frequency and voltages also in case the main grid support is missing. Future research directions may involve the use on historical data of renewable sources' production and loads' consumption for improving the control design and increasing the energy efficiency. Additional research effort may be also devoted to the design of multi-agent algorithms for generic problem structures, as microgrids may easily involve non-linear models, constraints and costs and this should not affect their efficient coordination in supporting the upcoming energy transition.

References

1. Kundur P, Balu NJ, Lauby MG (1994) Power system stability and control. McGraw-Hill, New York
2. Joos G, Ooi BT, McGillis D, Galiana FD, Marceau R (2000) The potential of distributed generation to provide ancillary services. In: 2000 IEEE power engineering society summer meeting, vol 3, pp 1762–1767
3. Hatziargyriou ND, Anastasiadis AG, Vasiljevska J, Tsikalakis AG (2009) Quantification of economic, environmental and operational benefits of microgrids. In: 2009 IEEE Bucharest PowerTech, pp 1–8
4. Yuen C, Oudalov A (2007) The feasibility and profitability of ancillary services provision from multi-microgrids. In: 2007 IEEE Lausanne Power Tech. IEEE, pp 598–603
5. Lasseter RH, Paigi P (2004) Microgrid: a conceptual solution. In: 2004 IEEE 35th annual power electronics specialists conference (IEEE Cat. No. 04CH37551), vol 6. IEEE, pp 4285–4290

6. La Bella A (2020) Optimization-based control of microgrids for ancillary services provision and islanded operation. Doctoral dissertation, Italy
7. La Bella A, Farina M, Sandroni C, Scattolini R (2018) Microgrids aggregation management providing ancillary services. In: 2018 European control conference (ECC). IEEE, pp 1136–1141
8. Poplavskaya K, De Vries L (2019) Distributed energy resources and the organized balancing market: a symbiosis yet? Case of three European balancing markets. Energy Policy 126:264–276
9. Boyd S, Parikh N, Chu E (2011) Distributed optimization and statistical learning via the alternating direction method of multipliers. Now Publishers Inc
10. Bertsekas DP (1997) Nonlinear programming. J Oper Res Soc 48(3):334–334
11. Molzahn DK, Dörfler F, Sandberg H, Low SH, Chakrabarti S, Baldick R, Lavaei J (2017) A survey of distributed optimization and control algorithms for electric power systems. IEEE Trans Smart Grid 8(6):2941–2962
12. La Bella A, Farina M, Sandroni C, Scattolini R (Nov, 2020) Design of aggregators for the day-ahead management of microgrids providing active and reactive power services. IEEE Trans Control Syst Technol 28(6):2616–2624, https://doi.org/10.1109/TCST.2019.2939992
13. Samad T, Koch E, Stluka P (2016) Automated demand response for smart buildings and microgrids: the state of the practice and research challenges. Proc IEEE 104(4):726–744
14. Kim SJ, Giannakis GB (2013) Scalable and robust demand response with mixed-integer constraints. IEEE Trans Smart Grid 4(4):2089–2099
15. Chang TH (2016) A proximal dual consensus ADMM method for multi-agent constrained optimization. IEEE Trans Signal Process 64(14):3719–3734
16. Justo JJ, Mwasilu F, Lee J, Jung JW (2013) AC-microgrids versus DC-microgrids with distributed energy resources: a review. Renew Sustain Energy Rev 24:387–405
17. La Bella A, Cominesi SR, Sandroni C, Scattolini R (2016) Hierarchical predictive control of microgrids in islanded operation. IEEE Trans Autom Sci Eng 14(2):536–546
18. La Bella A, Negri S, Scattolini R, Tironi E (2018) A two-layer control architecture for islanded AC microgrids with storage devices. In: 2018 IEEE conference on control technology and applications (CCTA), pp 1421–1426
19. La Bella A, Nahata P, Ferrari-Trecate G (2019) A supervisory control structure for voltage-controlled islanded DC microgrids. In: 2019 IEEE 58th conference on decision and control (CDC). IEEE, pp 6566–6571

Open Access This chapter is licensed under the terms of the Creative Commons Attribution 4.0 International License (http://creativecommons.org/licenses/by/4.0/), which permits use, sharing, adaptation, distribution and reproduction in any medium or format, as long as you give appropriate credit to the original author(s) and the source, provide a link to the Creative Commons license and indicate if changes were made.

The images or other third party material in this chapter are included in the chapter's Creative Commons license, unless indicated otherwise in a credit line to the material. If material is not included in the chapter's Creative Commons license and your intended use is not permitted by statutory regulation or exceeds the permitted use, you will need to obtain permission directly from the copyright holder.

Chapter 13
Allowing a Real Collaboration Between Humans and Robots

Andrea Casalino

13.1 Introduction

Up to the recent past, the paradigm universally adopted for industrial robotics provided for the strict segregation of robots in protected environments, adopting fences or optical barriers. Only recently, the potential benefits of a collaboration between humans and robots have gained the attention of roboticists [10], mainly motivated by the Industry 4.0 paradigm [15]. Since many tasks are still impossible to be fully automatized, it seems natural to let humans and robots cooperate: highly cognitive actions are undertaken by humans, while those requiring high precision and repeatability are performed by robots. A well established literature addressed the problem of a safe coexistence in a shared space by introducing motion control techniques [6], based on the use of sensors perceiving the scene and tracking the motion of the human operators. Corrective trajectories can be planned on-line with the aim of dodging the human and, at the same time, keep driving the manipulator to the desired target position.

Such initial works conceived the robots as something that should interfere as less as possible with the humans populating the same cell, only opening the door for a real collaboration. The possibility for a properly instrumented robotic device to understand and somehow predict humans' intentions is now considered as important as safety and it is possible by providing the robots with the proper cognitive capabilities. In this context, a crucial role is played by vision sensors since the analysis of human motion is one of the most important features. Many results have been reported showing the increasing capability of robots to semantically interpret their human fellows. In [11] a method based on conditional random files (CRF) is used by the robot to anticipate its assistance. In [12] Gaussian Mixture Models (GMMs) are used to predict human reaching targets.

A. Casalino (✉)

Dipartimento di Elettronica, Informazione e Bioingegneria, Politecnico di Milano, Piazza Leonardo Da Vinci 32, Milan, Italy

e-mail: andrea.casalino@polimi.it

© The Author(s) 2021

A. Geraci (ed.), *Special Topics in Information Technology*,

PoliMI SpringerBriefs, https://doi.org/10.1007/978-3-030-62476-7_13

Collaborative assemblies are typical applications of collaborative robotics. In such scenarios, the actions of the robots influence the ones of the humans, and the optimization of the robotic action sequence becomes crucial. Reference [4] describes a genetic algorithm for a collaborative assembly station which minimises the assembly time and costs. In [14], a trust-based dynamic subtask allocation strategy for manufacturing assembly processes has been presented. The method, which relies on a Model Predictive Control (MPC) scheme, accounts for human and robot performance levels, as well as far their bilateral trust dynamics. By taking inspiration from real-time processor scheduling policies, [9] developed a multi-agent task sequencer, where task specifications and constraints are solved using a MILP (Mixed Integer Linear Programming) algorithm, showing near-optimal task assignments and schedules. Finally, [13] proposes a task assignment method, based on the exploration of possible alternatives, that enables the dynamic scheduling of tasks to available resources between humans and robots. Many alternative plans are computed in [16], where a dynamic task allocation of activities is also possible, by taking into account the different capabilities of agents (robots and humans). In [3], the scheduling problem is solved using a Generalised Stochastic Petri Net as a modelling tool. The selection of the optimal plan takes into account the amount of time for which the agents remain inactive, waiting for the activation of some tasks.

All the aforementioned works seem to model the human operators as no more than highly cognitive manipulators, that can be instructed to do certain actions. In this work timed Petri nets will be adopted to model the human-robots multi agent system, however assuming the human as an uncontrollable agent, whose actions must be recognized in order to decide which complementary actions the robots should undertake. The remaining of this chapter is structured as follows: Sect. 13.2 will discuss the activity recognition problem, while Sect. 13.3 will present the approach followed to predict the future ones. Such predictions will be crucial for applying the scheduling approach described in Sect. 13.4.

13.2 Recognizing the Human Actions

Recognizing the actions performed by a human is easy and natural for another human being. It is done by mainly analysing the motion of the arms. This is the same approach followed by computer vision algorithms. In the following, an approach based on RGB-d cameras and factor graphs will be discussed. Assume the (finite) set of possible human actions $\mathcal{A} = \{a_1, \ldots, a_m\}$ known. The aim of the algorithm proposed in this section is to detect the starting and the ending time of each action, by analysing the motion performed by the operator in the recent past. RGB-D sensors can be exploited to keep track of some points of interest in the human silhouette, see the top left corner of Fig. 13.1. The signals retrieved are subdivided into many sub-windows each having a maximum length of l_w samples. Then, a feature vector $F_O^i = \begin{bmatrix} O_1^i \ldots O_F^i \end{bmatrix}$ made of F components can be extracted from the ith window, representing an indirect indication of the action $a \in \mathcal{A}$ that was performed within

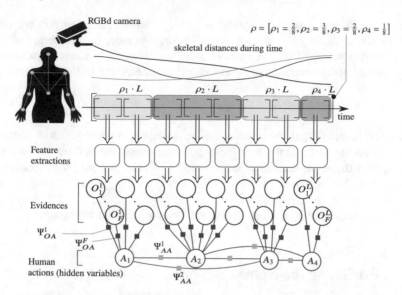

Fig. 13.1 A RGBd camera can be used to keep track of the skeletal points (cyan dots in the top left corner) and compute the skeletal distances over time. The segmenting graph is built considering a segmentation hypothesis ρ and the observations are partitioned accordingly

the same window. A good selection for the feature set is crucial and can be made with deep learning techniques for instance. Clearly, a specific action a must produce characteristic values for F_O. In the layout proposed in Sect. 13.4, the inter-skeletal distances as well as the distance of the operator's wrists to some particular positions of the space were empirically found to be effective.

The main problem to overcome when segmenting human actions is that the exact durations of the actions (and consequently the starting and ending time instants) as well as their total number is not precisely known. To handle this aspect, a probabilistic framework must be considered and the most probable sequence of actions that produced a certain macro-window of observations must be determined. This problem can be tackled considering factors graph [7], having as hidden variables the sequence of actions actually performed by the human and as evidences a battery of features $F_O^{1,...,L}$. Vector $\rho = \begin{bmatrix} \rho_1 \dots \rho_S \end{bmatrix}$ is adopted for describing the durations of human actions as well as their number. Indeed, ρ_i indicates the percentage of time spent by the human when doing the ith action in the sequence. Knowing ρ, it is possible to build the underlying factor graph, assuming to connect $\rho_i \cdot L$ observations to the node representing the ith action, refer to Fig. 13.1. The potentials Ψ_{OA} must express the correlation existing among the features and the actions, while those correlating the actions over time, Ψ_{AA}^i, should take into account precedence constraints among the actions or any kind of prior knowledge about the process. Since the real segmentation ρ^* describing the sequence of actions is not available, the proposed approach considers many hypotheses[1] $\rho^1, \rho^2, \ldots$, which are iteratively compared with the aim of finding the optimal one $\hat{\rho}$, i.e. the one more in accordance with the observations retrieved from the sensors. More specifically, a genetic algorithm [8] can be efficiently adopted assuming as a fitness function C the following likelihood:

$$C(\rho) = \mathcal{L}(\rho | O_1^1, \ldots, O_F^L) = \mathbb{P}(O_1^1, \ldots, O_F^L | \rho)\mathbb{P}(\rho)_{prior} \qquad (13.1)$$

where $\mathbb{P}(O_1^1, \ldots, O_F^L | \rho)$ can be computed by doing belief propagation on the factor graph pertaining to a specific hypothesis ρ. Determining the optimal segmentation $\hat{\rho}$ completes the segmentation task: the number of actions done by the operator is assumed equal to the number of segments in ρ and for the jth segment the marginal distribution $\left[\mathbb{P}(A_j = a_1 | O_1^1, \ldots, O_F^L, \rho^*) \ldots \mathbb{P}(A_j = a_m | O_1^1, \ldots, O_F^L, \rho^*)\right]$ expresses in a probabilistic way which action was done within that segment.

13.3 Predicting the Human Actions

By knowing the sequence of actions done in the recent past, predictions about the future ones can be made. In particular, we are interested in evaluating the waiting time τ^a before seeing again a particular action $a \in \mathcal{A}$.

Human assembly sequences usually form quasi-repetitive patterns. In other words, the sequence of human activities can be modelled through a time series, which is the output of a certain dynamic process. Assume A_k as the ongoing activity at discrete time instant k, the behaviour of the human fellow co-worker can be modelled through the following discrete-time process:

$$A_{k+1} = f(A_k, A_{k-1}, A_{k-2}, \ldots, A_{k-n})$$
$$t_{k+1} = t_k + g(A_k) \qquad (13.2)$$

where $t_k \in \mathbb{R}^+ \cup \{0\}$ represents the time instant corresponding to the transition from A_{k-1} to A_k and $g(a) = T^a > 0$ is the duration of activity $a \in \mathcal{A}$. The stochasticity of the underlying discrete process governing the sequence of activities can be modelled by making use of a higher order Markov Model [5], which computes the predicting probability distribution associated to the next activity in the following way:

$$\mathbb{P}(A_{k+1} = a | A_k = k_0, \ldots, A_{k-n} = k_n) \approx$$
$$\approx \sum_{i=0}^{n} \lambda_i \mathbb{P}(A_{k+1} = a | A_{k-i} = k_i) \qquad (13.3)$$

---

[1]Notice that also the number of elements in ρ is not known and many hypothesis with a different number of actions should be considered.

hence a mixture model that corresponds to usual Markov Chains for $n = 0$ and requires only $m^2 (n + 1)$ parameters. A prediction of the probability distribution $\hat{X}_{k+1}$ at time $k + 1$ can be computed as

$$\hat{X}_{k+1} = \sum_{i=0}^{n} \lambda_i Q_i X_{k-i} \tag{13.4}$$

where $X_{k+1} = \left[\mathbb{P}(A_{k+1} = a_1) \ldots \mathbb{P}(A_{k+1} = a_m) \right]^{T}$ refers to the probability distribution describing the state of the system[2] at step $k + 1$, while Q_i denotes the i-steps transition probability matrix that can be simply evaluated through counting statistics. The weights λ_i are estimated on-line, by minimizing the prediction errors on a window of n recent observations:

$$min \left(\left\| \sum_{i=0}^{n} \lambda_i Q_i X_{k-i} - X_{k+1} \right\|^{2} \right) \tag{13.5}$$

It is not difficult to show that this leads to the solution of a quadratic programming problem in the following form:

$$\min_{\lambda} \| A\lambda - b \|^2 \text{ subject to } \sum_{i=0}^{n} \lambda_i = 1, \text{ and } \lambda_i \geq 0 \tag{13.6}$$

Assume that the duration of an activity $a \in \mathcal{A}$, i.e. T^a, can be modelled as a stochastic variable with a strictly positive lower bound, i.e. $T^a \geq T^a_{min} > 0$. In order to estimate the waiting time needed for the certain activity a to show up, say τ^a, we can combine this information with the model describing the activity sequence, Eq. (13.3). In particular, at the present continuous time instant $\bar{t}$, given the sequence of the last activities (possibly including the currently running one) $A_k, A_{k-1}, \ldots, A_{k-n}$, we would like to estimate the probability distribution of the waiting time for the beginning of a certain activity a, i.e. $\mathbb{P}(\tau^a \leq t \,|\, A_k, A_{k-1}, \ldots, A_{k-n})$. The key idea is to construct a predictive reachability tree. Then, evaluating the time spent to traverse each possible branch in the tree, terminating with the desired activity $a \in \mathcal{A}$, it is possible to estimate τ^a. Since the reachability tree is, in principle, infinite, a prediction horizon ΔT must be defined, meaning that the given probability will be computed up to the instant $t = \bar{t} + \Delta T$. The probability associated to each branch can be simply computed by multiplying the probability of each arc of the branch, i.e. $p_{branch} = \prod_{(i,j) \in branch} p_{(i,j)}$. As for the waiting time associated to each branch τ_{branch}, this is simply the sum of the duration of each activity T^a, i.e. $\tau_{branch} = \sum_{j:(i,j) \in branch} T^j$. The time associated to each branch is computed as the sum of stochastic variables which are empirically approximated using the statistics associated to recently acquired samples, using a Monte Carlo numerical approach.

---

[2]Which is exactly the output of the algorithm described in Sect. 13.2.

Finally, given the distributions of the times associated to each branch, the overall distribution of the waiting time of the activity a can be simply computed as a weighted sum of the waiting times associated to each branch, i.e.

$$
\mathbb{P}\left(\tau^a \leq t \,|\, A_k, A_{k-1}, \ldots, A_{k-n}\right) = \\
= \sum_{branch} p_{branch} \mathbb{P}\left(\tau_{branch} \leq t\right). \tag{13.7}
$$

13.4 Assistive Scheduling

Detecting and predicting the human actions is crucial when planning the robotic ones. Assuming the tasks as pre-assigned to all the agents (robots and humans), the aim of scheduling become essentially to control the robots in a way as much as possible compliant with the human (predicted) intentions. To this purpose the cell is modelled as a multi agent system by making use of timed Petri Nets [1]. In particular, the robots will be modelled as usual controllable agents, while humans will be assumed as uncontrollable, even if they will affect the system when doing the assigned tasks. Modelling Timed Petri Nets can be done by following systematic rules, from a description of the tasks assigned to both humans and robots as well as their precedence constraints, Figs. 13.2 and 13.3. The robotic tasks are modelled in a canonical way with two transitions: the first starting the task and the second terminating it. For every robots, an additional action is always considered and consists of an idle of a quantum of time (orange transitions in Fig. 13.3), able to postpone the starting of the following action (multiple idle are also possible). The human actions are modelled with three transitions: the first one firing when the human is predicted to be ready to start the corresponding action (also indicated as intentional transition), the second one firing as soon as the human actually starts that action and the last one terminating it. Red places in Fig. 13.3 are those related to idle states for the agents.

Scheduling essentially consists in deciding the commands to dispatch to the robots, or in other words resolving the controllable conflicts. To this purpose, a model pre-

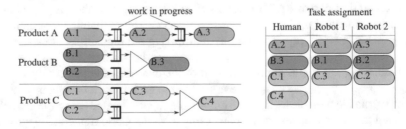

Fig. 13.2 Example of task allocation of some assembly sequences

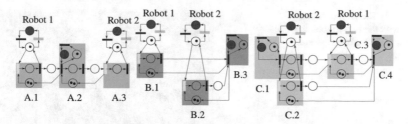

Fig. 13.3 The timed Petri nets adopted to model the co-assemblies reported in Fig. 13.2. The global net modelling the system can be obtained by superimposing the reported three ones. The transitions reported in yellow are those assumed as controllable. Orange transitions are controllable too and correspond to idling of a quantum of time. Red places are those related to idle state for an agent of the system

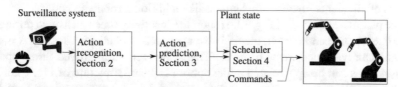

Fig. 13.4 The assistive scheduling approach

dictive approach is exploited and the underlying Petri Net is used to compare possible future evolutions of the systems with the aim of finding those minimizing the inactivity times and indirectly maximizing the throughput. Since the system is only partially controllable, a stochastic approach is assumed and the optimal policy is recomputed every time by exploring the reachability tree of the net with a Monte Carlo simulation, details are explained in [2]. The predictions about the future human activities, Sect. 13.3, are used to build the distributions modelling the intentional transitions. The pipeline of Fig. 13.4 summarizes the entire approach.

13.5 Results

The assistive scheduling approach described so far was applied in a realistic co-assembly, whose workspace consisted in: one IRB140 ABB robot, the collaborative dual-arm robot YuMi of ABB, two automatic positioners (treated like additional robots to control) and a MICROSOFT KINECT v2 as monitoring device, see Fig. 13.5. The aim of the collaboration was to perform the assembly of a box containing a USB pen.[3] 20 volunteers were enrolled for the experiments and were divided into two equally sized groups named Group 1.A and Group 1.B. Operators in Group 1.A performed the assembly while the strategy discussed in Sect. 13.4 was adopted.

---

[3]The assembly steps are described at https://youtu.be/_Jxo1mNZ1C8.

Fig. 13.5 The experimental
setup with the two robots, the
carts and the human operator

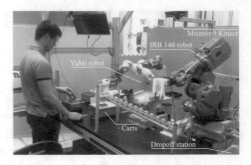

Instead, for those in Group 1.B a centralized strategy was adopted for deciding the actions of all agents, treating the humans like additional robots to control.

The pictures at the top of Fig. 13.6 report the inactivity times measured during the experiments. As can be seen, the assistive approach applied for Group 1.A is able to significantly reduce waste of times. This reflects in an improved throughput of the system as can be appreciated in the bottom left corner picture of Fig. 13.6. It is worth to recall that the approach followed to predict the human intentions, Sect. 13.3 is data-driven, adapting to the human over time. This fact gives the scheduler some learning capabilities or in other words the produced plans are able to constantly improve the way the robots are controlled, refer to the bottom right corner of Fig. 13.6.

13.6 Conclusions

This work aimed to study how to allow humans and robots to actively collaborate for performing a shared tasks, like the ones involved in co-assemblies. Humans were not conceived as additional controllable agents which are instructed to perform actions scheduled by a centralized planner. Instead, the decision-making capabilities of the operators were exploited, since they were allowed to drive the evolution of the plant. In this context, the robotic mates were endowed with the cognitive capabilities required for both recognizing and predicting the human actions. Regarding action recognition, particular attention was paid to manage the fact that the starting and ending time instants of the action were not known. To overcome such issue, genetic algorithms were deployed in conjunction with factor graphs. For what concerns the prediction problem, a data-driven approach was described, using an adaptive Higher order Markov model to forecast the sequence of human actions. The scheduling of the robotic actions was done with a kind of scenario based approach, which needs to explore the reachability tree of a timed Petri Net modelling the agents in the robotic cell.

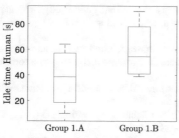

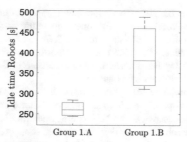

(a) Distributions of idling times for the human

(b) Distributions of total idling times for the robots

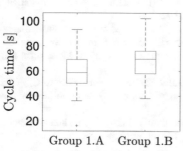

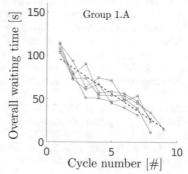

(c) Distributions of the cycle times

(d) The evolution of the overall wait time of the agents (the idling times of all the agents were summed), during the experiments of Group 1.A. The dashed red curve is the regressed line interpolating all the data.

Fig. 13.6 Results showing the benefits of an assistive scheduling, Group 1.A w.r.t to a centralized approach, Group 1.B

References

1. Berthomieu B, Diaz M (1991) Modeling and verification of time dependent systems using time Petri nets. IEEE Trans Softw Eng 17(3):259–273
2. Casalino A, Zanchettin A, Piroddi L, Rocco P (2019) Optimal scheduling of human-robot collaborative assembly operations with time Petri nets. IEEE Trans Autom Sci Eng
3. Chen F, Sekiyama K, Huang J, Sun B, Sasaki H, Fukuda T (2011) An assembly strategy scheduling method for human and robot coordinated cell manufacturing. Int J Intell Comput Cybern 4:487–510
4. Chen F, Sekiyama K, Cannella F, Fukuda T (2014) Optimal subtask allocation for human and robot collaboration within hybrid assembly system. IEEE Trans Autom Sci Eng 11(4):1065–1075
5. Ching WK, Fung ES, Ng MK (2004) Higher-order Markov chain models for categorical data sequences. Nav Res Logist (NRL) 51(4):557–574
6. Flacco F, Kroger T, De Luca A, Khatib O (2012) A depth space approach to human-robot collision avoidance. In: 2012 IEEE international conference on robotics and automation, pp 338–345

7. Frey BJ (2002) Extending factor graphs so as to unify directed and undirected graphical models. In: Proceedings of the nineteenth conference on uncertainty in artificial intelligence. Morgan Kaufmann Publishers Inc., pp 257–264
8. Goldberg DE (1989) Genetic algorithms in search, optimization, and machine learning, Addison Wesley, Reading, MA. The applications of GA-genetic algorithm for dealing with some optimal calculations in economics
9. Gombolay MC, Wilcox R, Shah JA (2013) Fast scheduling of multi-robot teams with temporospatial constraints. Robotics: science and systems
10. Haddadin S, Suppa M, Fuchs S, Bodenmüller T, Albu-Schäffer A, Hirzinger G (2011) Towards the robotic co-worker. Springer, Berlin, pp 261–282
11. Koppula HS, Saxena A (2016) Anticipating human activities using object affordances for reactive robotic response. IEEE Trans Pattern Anal Mach Intell 38(1):14–29
12. Luo R, Hayne R, Berenson D (2017) Unsupervised early prediction of human reaching for human–robot collaboration in shared workspaces. Auton Robot 1–18
13. Nikolakis N, Kousi N, Michalos G, Makris S (2018) Dynamic scheduling of shared human-robot manufacturing operations. Procedia CIRP 72:9–14
14. Rahman SMM, Sadrfaridpour B, Wang Y (2015) Trust-based optimal subtask allocation and model predictive control for human-robot collaborative assembly in manufacturing. In: ASME 2015 dynamic systems and control conference. American Society of Mechanical Engineers, pp V002T32A004–V002T32A004
15. Robla-Gómez S, Becerra VM, Llata JR, González-Sarabia E, Torre-Ferrero C, Pérez-Oria J (2017) Working together: a review on safe human-robot collaboration in industrial environments. IEEE Access 5:26754–26773
16. Tsarouchi P, Matthaiakis A-S, Makris S, Chryssolouris G (2017) On a human-robot collaboration in an assembly cell. Int J Comput Integr Manuf 30(6):580–589

Open Access This chapter is licensed under the terms of the Creative Commons Attribution 4.0 International License (http://creativecommons.org/licenses/by/4.0/), which permits use, sharing, adaptation, distribution and reproduction in any medium or format, as long as you give appropriate credit to the original author(s) and the source, provide a link to the Creative Commons license and indicate if changes were made.

The images or other third party material in this chapter are included in the chapter's Creative Commons license, unless indicated otherwise in a credit line to the material. If material is not included in the chapter's Creative Commons license and your intended use is not permitted by statutory regulation or exceeds the permitted use, you will need to obtain permission directly from the copyright holder.

Printed in the United States
By Bookmasters